This dissertation presents the research I conducted as part of the final requirement for a Ph.D. in the Electrical Engineering Department at the University of Pittsburgh. The dissertation was completed in 1991. Part of the research had already been published in an article titled, "Superconducting Synchronous Induction Motor Performance," in the peer-reviewed IEE PROCEEDINGS, Vol. 137, Pt. B, No. 2, March 1990, pp 120-123. Unfortunately, my dissertation was not accepted at the University of Pittsburgh so the Ph.D. was not granted.

After leaving the University of Pittsburgh, I sent a copy of the dissertation to UMI/Bell & Howell Information and Learning. This company maintained a database of works like my dissertation. I had hoped that my dissertation would be available to whoever wished to see it. Recently, I made a phone call to the company that took over the UMI/Bell & Howell files. They said that they no longer had my dissertation in their files.

Following is a scan of a bound hardcover copy of the dissertation. Thanks to Kindle Direct Publishing, KDP, it is available to you.

Sincerely,

6/29/2025

DESIGN AND ANALYSIS OF A SUPERCONDUCTING HIGH SPEED SYNCHRONOUS/INDUCTION MOTOR

by

Stephen Philip Tubbs

B.S. in Applied Mathematics and Engineering Physics,

University of Wisconsin

M.Sc. in Electrical Engineering,

Heriot-Watt University

©Registration Number TX3134696

Effective Date of Registration 8/14/1991

ISBN 9780965944632

1344 Firwood Dr.
Pittsburgh, PA 15243

Signed

FOREWORD

This dissertation was created to satisfy part of the requirements for the Ph.D. in the electrical engineering department of the University of Pittsburgh, Pittsburgh, PA. Unfortunately, it was not accepted there.

Thanks to the thoughtfulness of your librarian this dissertation has been placed here. It is hoped that future electric motor researchers will find it useful in their own motor research.

I would like to thank my wife Jenann for her encouragement. Thanks go to Ralph Erhard of Westinghouse (ret.), Edward Harbist of AEG/Westinghouse Transportation, Prof. James L. Kirtley of the Massachusetts Institute of Technology, Dr. Stephen Kuznetsov of the Electric Machinery Associates, Prof. Jay A. Switzer of the Univ. of Missouri-Rolla and Eugene Whitney of Westinghouse (ret.). Thanks go to the Magnetic Metals Corporation for their donations of motor and stator laminations that were invaluable to the research. Thanks also go to John Badura and the other machinists in the University of Pittsburgh, Benedum Hall machine shop who helped me in the construction of precision motor parts.

ABSTRACT

Signature________________

DESIGN AND ANALYSIS OF A SUPERCONDUCTING HIGH SPEED SYNCHRONOUS/INDUCTION MOTOR

Various motors were designed, built, analyzed and/or tested.

Using standard motor design procedures a high T_c (also called high temperature, ceramic or liquid nitrogen temperature) superconducting synchronous/induction motor was designed. This design was created as a modification of an existing 1,200 rpm, 15 hp induction motor design. The design was then tested analytically with computer simulations. A design was also considered for a high T_c superconducting high speed synchronous/induction motor.

Several miniature high speed induction motors were designed, built, and then tested at liquid nitrogen temperatures. They were prototypes for high T_c superconducting high speed synchronous/induction motors.

None of the prototype motors operated in the superconducting state as a true superconducting motor.

It is concluded that superconducting high speed synchronous/induction motors using high T_c superconductors in their rotors will be practical only when better superconductors are developed.

There is room for future research. These motors have several major problems that must be overcome: 1) Presently, high T_c superconductor materials can not withstand high enough current or magnetic field densities of the required frequencies. 2) Present high T_c superconductor materials are fragile and impossible to join or weld together into superconducting junctions.

The motors also have several disadvantages relative to ordinary motors. 1) A refrigeration system is necessary to maintain superconductivity., 2) Standard bearings and lubricants do not perform well at liquid nitrogen temperatures., 3) There will be serious torque, speed, and current oscillations when the superconducting material makes the transition to superconductivity or a sudden load is applied.

DESCRIPTORS

High Speed Motor

Superconducting Motor

Synchronous Motor

Induction Motor

Superconductor Application

Synchronous/Induction Motor

TABLE OF CONTENTS

LIST OF FIGURES

NOMENCLATURE

A_b = cross-sectional area of the superconducting portion of a rotor bar (m^2)

A_{sgrl} = angle between the stator phase current and the voltage induced in the stator by the rotating flux (rad)

$B_{ab}(t)$ = flux density being cut by the a phase rotor bar as a function of time (T)

$B_{bb}(t)$ = flux density being cut by the b phase rotor bar as a function of time (T)

$B_{cb}(t)$ = flux density being cut by the c phase rotor bar as a function of time (T)

B_r = flux density through the outer portion of a superconducting rotor bar (T)

C_n = heat of vaporization of liquid nitrogen (J/kg)

C_{orf} = correction factor to account for stator and rotor lamination reluctance

D = inside diameter of stator (m)

E = induction motor equivalent circuit phasor voltage across R_m and X_m (peak V)

$E(t_s)$ = induction motor equivalent circuit phasor voltage across R_m and X_m immediately prior to the transition to superconductivity (peak V)

E_{ff} = motor efficiency (pu)

$E_{ff}(t)$ = efficiency as a function of time (per unit)

E_r = voltage generated per perfect pitch rotor coil (peak V)

E_T = magnitude of stator line to neutral input voltage (peak V)

E_T = phasor value of stator line to neutral input voltage (peak V)

$E_{T\phi}$ = magnitude of line to neutral voltage induced by the rotating flux onto each stator phase winding (peak V)

$\mathbf{E}_{Ts}$ = phasor value of the line to neutral voltage induced by the rotating flux onto each stator phase winding (peak V)

$i_{ab}(t)$= a phase rotor bar current as a function of time (A)

$i_{ar}(t)$= two axis theory direct axis secondary (rotor) current as a function of time (A)

$i_{as}(t)$= two axis theory direct axis primary (stator) current as a function of time (A)

$i_{bb}(t)$= b phase rotor bar current as a function of time (A)

$i_{br}(t)$= two axis theory quadrature axis secondary (rotor) current as a function of time (A)

$i_{bs}(t)$= two axis theory quadrature axis primary (stator) current as a function of time (A)

$i_{cb}(t)$= c phase rotor bar current as a function of time (A)

I = magnitude of stator phase current (peak A)

$\mathbf{I}$ = phasor value of stator phase current (peak A)

I_m = magnitude of stator magnetizing current (peak A)

$I_r(\theta_s)$= magnitude of rotor bar current as a function of angle around the stator (A)

I_{rb} = magnitude of rotor bar current (peak A)

I_{rbl} = maximum magnitude of rotor bar current immediately prior to the transition to superconductivity (A)

$\mathbf{I}_{re}$ = phasor value of induction motor equivalent circuit current in the rotor section (peak A)

$I_{rs}(\theta_s)$= magnitude of rotor bar current to oppose the effect of stator current after transition as stator current and power angle vary. It is as a function of angle around the stator. (A)

J = rotor and load moment of inertia (kg m^2)

J_r = rotor bar current density (peak A/m^2)

k_d = stator winding distribution factor

k_p = stator winding pitch factor

k_r = air gap coefficient for rotor slots

k_s = air gap coefficient for stator slots

K_w = maximum speed-power constant

l_g = rotor stack height (m)

L_{lr} = two axis theory secondary (rotor) leakage inductance, same for direct and quadrature axis (H)

L_{ls} = two axis theory primary (stator) leakage inductance, same for direct and quadrature axis (H)

L_{re} = induction motor equivalent circuit rotor portion inductance (H)

L_{rb} = rotor bar and adjacent end-ring inductance (H)

L_s = stator inductance per phase and induction motor equivalent circuit stator portion inductance (H)

m = number of phases

M = maximum magnitude two axis theory mutual inductance between stator and rotor, same for direct and quadrature axes (H)

N = number of stator conductors in series per phase

N_b = number of rotor bars per phase group

p = number of poles

$P_{inA}(t)$= power input to the A phase of stator as a function of time (W)

$P_{av}(t)$= average motor output power as a function of time (watts)

P_e = total electrical motor power produced (W)

PF = motor power factor (pu)

P_{fl} = fixed losses (W)

P_{fws} = friction and windage power loss at synchronous rotor speed (W)

P_{in} = total input power to motor (W)

$P_{in}(t)$= input power to the motor as a function of time (watts)

$PF(t)$= power factor of power to the motor as a function of time (watts)

P_{Mag} = magnetic power lost in motor (W)

P_{Mout} = maximum speed-power motor power capacity (kw)

P_{out} = motor mechanical output power (W)

P_{rt} = total power going from the stator to the rotor (W)

P_{Rs} = stator resistive power loss (W)

r = distance from the rotor center to the midpoint of the rotor bars (m)

r_{ar} = two axis theory direct axis secondary (rotor) resistive impedance (ohms)

r_{as} = two axis theory direct axis primary (stator) resistive impedance (ohms)

r_{br} = two axis theory quadrature axis secondary (rotor) resistive impedance (ohms)

r_{bs} = two axis theory quadrature axis primary (stator) resistive impedance (ohms)

R_c = a corrective resistance value that approximates the effective magnetizing power loss in the synchronous motor equivalent circuit (Ω)

R_{LN} = cost of liquid nitrogen ($/kg)

$R_{LN/CV}$ = ratio of effective power to create liquid nitrogen to power cooling value of vaporizing liquid nitrogen

R_M = induction motor equivalent circuit magnetizing resistance (Ω)

R_p = commercial electricity rate ($/J)

R_{rb} = rotor bar and adjacent end-ring resistance (Ω)

R_{re} = induction motor equivalent circuit rotor portion resistance (Ω)

R_s = stator resistance per phase and induction motor equivalent circuit stator portion resistance (Ω)

R_{sy} = synchronous motor equivalent circuit resistance (Ω)

R_{t1} = resistance of copper at temperature t_1 (Ω)

R_{t2} = resistance of copper at temperature t_2 (Ω)

s = induction motor rotor slip (pu)

s_{wr} = width of a rotor slot (m)

t = time (s)

t_s = time of transition to superconductivity (s)

t_1 = initial temperature of copper ($^\circ$C)

t_2 = final temperature of copper ($^\circ$C)

$T_{av}(t)$ = average rotor torque to load as a function of time (n m)

T_c = critical temperature at which a material becomes superconductive ($^\circ$C)

T_e = total electrical rotor torque (N m)

$T_e(t)$ = total electrical rotor torque as a function of time (N m)

T_g = electrical torque produced by one symmetrically displaced group of rotor bar pairs (N m)

T_l = load torque (N m)

T_{out} = motor mechanical output torque (N m)

$T_{out}(t)$ = rotor torque to load as a function of time (n m)

$T(t)$ = output torque of the motor as a function of time (N m)

$v_{ds}(t)$ = two axis theory direct axis applied (stator) voltage as a function of time, a two axis transformation of the per phase voltage applied in previous 3 phase analysis (V)

$v_{bs}(t)$= two axis theory quadrature axis applied (stator) voltage as a function of time, a two axis transformation of the per phase voltage applied in previous 3 phase analysis (V)

V_{as} = two axis theory two axis theory magnitude of direct axis applied (stator) voltage, a two axis transformation of the per phase voltage applied in previous 3 phase analysis (peak V)

V_{bs} = two axis theory magnitude of quadrature axis applied (stator) voltage, a two axis transformation of the per phase voltage applied in previous 3 phase analysis (peak V)

X_M = induction motor equivalent circuit magnetizing inductive impedance (Ω)

X_{re} = induction motor equivalent circuit rotor portion inductive impedance (Ω)

X_s = stator inductive impedance per phase and induction motor equivalent circuit stator portion inductive impedance (Ω)

X_{sy} = synchronous motor equivalent circuit inductive impedance (Ω)

δ = rotor to stator radial air gap (m)

δ_p = angle between the voltage generated on the stator phase winding by the rotating flux and the stator input voltage (rad)

$\theta_r(t)$ = two axis theory angular position of the rotor as a function of time (radians)

θ_s = angle measured around the stator from the stator's frame of reference (rad)

$\phi(t)$ = flux through a one turn loop as a function of time (Wb)

$\phi_{dr}(t)$= two axis theory direct axis flux through the rotor times the number of turns in the rotor direct axis as a function of time (Wb)

$\phi_{as}(t)$= two axis theory direct axis flux through the stator times the number of turns in the stator direct axis as a function of time (Wb)

$\phi_{br}(t)$ = two axis theory quadrature axis flux through the rotor times the number of turns in the rotor quadrature axis as a function of time (Wb)

$\phi_{bs}(t)$ = two axis theory quadrature axis flux through the stator times the number of turns in the stator quadrature axis as a function of time (Wb)

ϕ_* = magnitude of flux per pole (Wb)

μ_o = magnetic permeability of free space, $4\pi \times 10^{-7}$ (T m/A)

w_* = maximum speed-power motor speed of revolution (rpm)

$w_r(t)$ = rotor speed as a function of time (rad/s)

$w_r(t_o)$ = rotor speed immediately prior to the transition to superconductivity (rad/s)

$w_{rav}(t)$ = average rotor speed as a function of time (rad/s)

w_s = synchronous rotor speed and angular frequency of the input voltage (rad/s)

1.0 INTRODUCTION

A superconducting synchronous/induction motor has a rotor winding that is capable of superconductivity. It will operate as an induction motor only during starting, before its windings become superconducting. Once superconducting, its rotor winding will not allow any more or less magnetic flux to pass through it than immediately before it became superconducting. The phrase "superconducting synchronous/induction motor" describes a motor that initially starts as an induction motor. During starting the rotor winding makes a transition from normal conductivity to superconductivity causing flux to become "trapped" in place on the rotor. After that, the motor runs similarly to a nonsalient pole permanent magnet synchronous motor.

The importance of the rotor flux being "trapped" must be emphasized. Once the rotor windings become superconducting, stator currents can not induce any more or less flux into the rotor. Because rotor flux is "trapped", classical motor analysis that assumes a sinusoidally time and position varying flux through the rotor can not be used. During superconducting operation the stator currents will induce discontinuous square wave like space waves of currents on to the rotor bars.

For a superconducting synchronous/induction motor to start as an induction motor it must have the ferromagnetic rotor structure of an induction motor. This structure must be used to help achieve the proper rotor resistance to inductance ratio for induction starting. However, the use of ferromagnetic materials also introduces the possibility of saturation. A superconducting rotor using standard laminations has a maximum magnetic flux density of about 2 T. This helps place a ceiling on the output power capacity of a superconducting synchronous/induction motor since power is proportional to flux.

A superconducting synchronous/induction motor also needs rotor resistance to start as an induction motor. This can be accomplished by forcing the material of the superconducting portion of the squirrel cage out of superconduction during initial starting. The rotors of the motors designed and analyzed here have copper squirrel cages next to and in parallel with a superconducting squirrel cage, as shown in Figure 1. The outer 1/6th of each rotor bar and end-ring is copper, the inner 5/6th is superconductor material. During initial starting the superconducting squirrel cage is kept out of superconductivity by not cooling it. Shortly after induction motor starting when the motor has nearly reached steady state, liquid nitrogen coolant is pumped through the cooling channels near the superconductors. When the temperature drops low enough the rotor currents suddenly shift from the parallel resistive copper squirrel cage to the zero resistance superconducting squirrel cage.

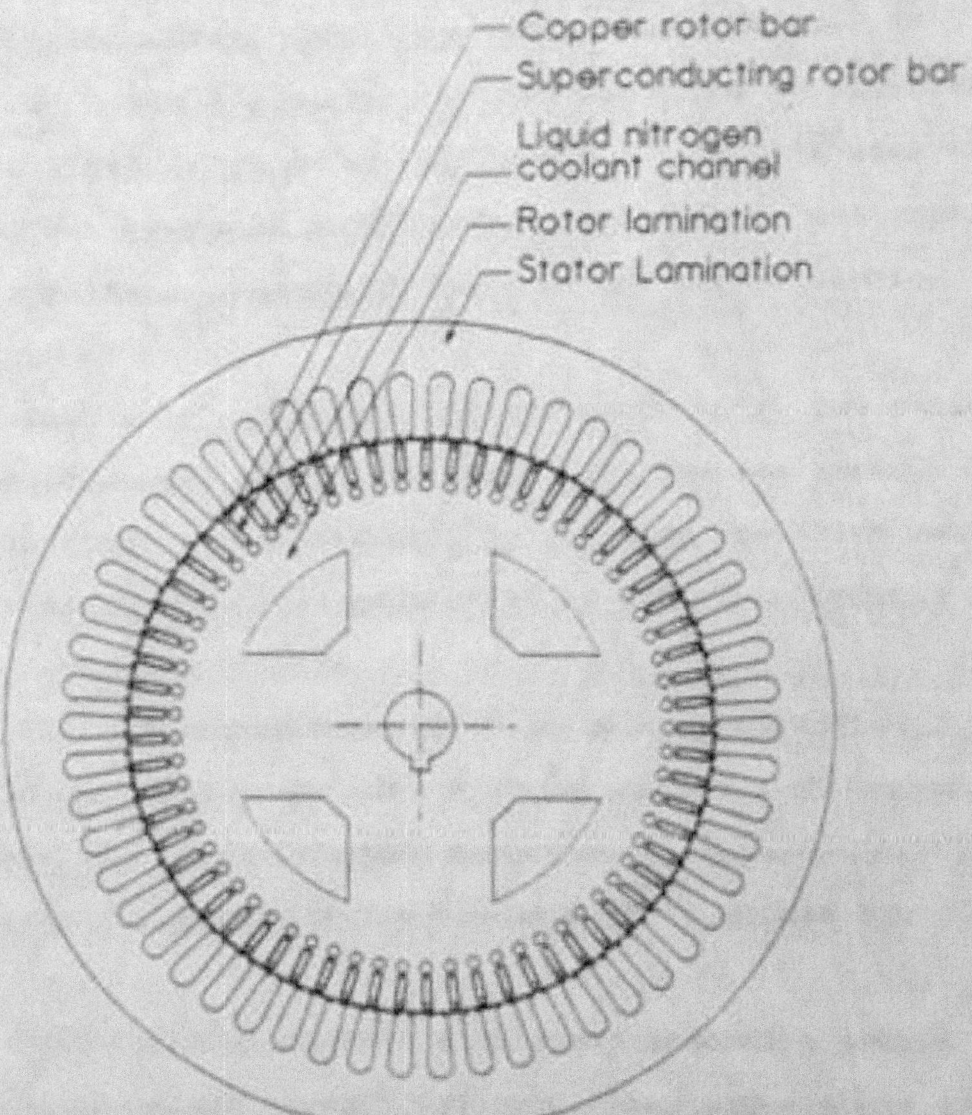

Figure 1 Sketch of part of a superconducting synchronous/induction motor cross-section. Stator conductors are not shown.

Superconducting synchronous/induction motors self-start as induction motors and then run as synchronous motors. They are more efficient than induction motors, if their effective cooling system power is less than the power saved by superconductive operation. They could be used on spacecraft where a cooling system might not be required or on extremely large power applications. In both uses they may offer increased efficiency, power factor, and capacity.

The use of superconductors in the stator winding is not considered.

The use of cryogenic, normal conducting, low resistance windings in the stator and rotor will improve induction motor operating efficiencies by reducing resistive heating. Induction motors have been built by others to operate at cryogenic temperatures.

The word "superconductor" in this dissertation refers to the relatively new high T_c (also called high temperature, ceramic or liquid nitrogen temperature) superconductors that have been developed since February 1987, unless otherwise indicated.

In this dissertation high speed induction motors have operating speeds above 9,000 rpm. They are similar to conventional lower speed induction motors that run below 3,600 rpm. The major differences between high speed and conventional lower speed motors are in bearings, output rating, and rotor diameter. Special bearings are required to withstand the higher speeds. Generally, high speed motors are more limited in their output power. To reduce vibration and centrifugal forces on outer rotor parts, high speed motors tend to have smaller rotor diameters.

1.1 Statement of the Problem

The purpose of this research is to design and analyze a motor that starts as an induction motor and runs at high speed as a superconducting synchronous motor.

1.2 Method of Approach

The approach to the design of a superconducting high speed synchronous/induction motor was to first research relevant published articles. Next, small experimental motors were built to run at low temperatures and high speeds. Using knowledge gained from these an existing normal induction motor design was modified to be a normal speed superconducting synchronous/induction motor. The approach to the analysis was confined to analysis of a normal speed superconducting synchronous/induction motor. Three different types of motor analysis were applied to the superconducting synchronous/induction motor design.

An intensive search for relevant articles was done manually and by an INSPEC data bank title and author search.

Miniature high speed induction motors (three stators and twelve rotors) were designed, built, and then tested at room and liquid nitrogen temperatures. They revealed problems that might have remained hidden if a purely analytical approach had been taken in the motor design.

A superconducting synchronous/induction motor design was created as a modification of an existing 1,200 rpm, 15 hp induction motor design.

Three different types of motor analysis were applied to this motor. The methods of motor analysis are as follows: the fixed sinusoidal rotating field method, the equivalent circuit method, and the two axis method. Each method of analysis produced its own system of equations. Then each system of equations was solved using computer programs.

Each method was applied to an ordinary induction motor (the state of the superconducting synchronous/induction motor before superconductivity), a special synchronous motor (the state of the superconducting synchronous/induction motor when superconductive), and the superconducting synchronous/induction motor in transition from normal conductivity to superconductivity.

1.3 Statement of Originality

To the best of my knowledge the approach taken in the analytical portion of this work is a significant addition to body of electrical engineering knowledge.

Prior to this work, there were no published methods of analysis of superconducting synchronous/induction motor performance. Also, no publication produced numerical predictions of motor efficiency, power factor and power output of a superconducting synchronous/induction motor. Prior to this work, there was no publication mentioning the substantial squirrel cage alternating currents that are induced into the squirrel cage of a superconducting synchronous/induction motor in steady state operation. The methods of this work predict numerical values for these alternating currents.

This dissertation is put forward as an original contribution to the field of electrical machinery research.

1.4 Need for Superconducting Synchronous/Induction Motors

Higher hp induction motors are used in industry and power generating stations. Efficiencies of these motors are on the order of 94%[(1)*]. If cooling costs are ignored, replacing these motors with superconducting synchronous/induction motors would reduce their losses by up to half. Power savings could be substantial. Furthermore superconducting synchronous/induction motors would be smaller and lighter, easing space and mounting requirements.

In spacecraft, actuator motors have a limited power supply and motor weight and size must be kept to a minimum. Superconducting synchronous/induction motors are more compact. Furthermore the cold environment of space means a pumped coolant system may not be necessary. A low T_c superconducting stepper motor has already been designed for space use as reported by Bugeat et al.[(2)].

1.5 Need for High Speed Motors

Motor power capacity is directly proportional to motor speed[(3)]. As power electronic circuits producing frequencies higher than 60 Hz continue to drop in cost some motor applications will benefit from the use of higher speed induction motors. Applications where weight and space must be conserved may justify the extra electronic circuitry necessary. 400 Hz motors have been used on aircraft for

---

*Parenthetical references placed superior to the line of text refer to the bibliography.

some time to conserve weight and space.

High speed motors are also useful in some machining applications such as grinding.

1.6 Need for Superconducting High Speed Synchronous/Induction Motors

A superconducting high speed synchronous/induction motor would offer compactness, power, and efficiency. It would be useful on spacecraft and in some ordinary large scale applications such as with boiler feed pumps and compressors. It might also be useful in tight enclosures such as with submerged pumps.

1.7 Literature Review

The field of rotating electrical machinery is one of the oldest in modern electrical engineering. The modern three phase induction motor was developed over one hundred years ago by Tesla. AC synchronous machines were developed before that. Because of the amount of research time and publication others have put into these topics a complete literature review is not a practical possibility. The literature review presented here covers the relatively recent developments reported during the last 40 years. The search was done by usual library methods and by an INSPEC data bank title and author search. The INSPEC search considered journals and conference reports that were written or abstracted in English since 1969.

1.7.1 Superconducting Synchronous/Induction Motors

The first paper considering superconducting synchronous/induction motors was by Leonhard[4]. In 1969 Leonhard showed theoretically that an induction machine with a superconducting winding will operate similarly to a synchronous machine. He also determined that the maximum torque output of such a superconducting machine is twice that of the breakdown torque of its normal conducting machine equivalent.

In 1979 Brechna and Kronig[5] reported a design for a 1.5 Mw three phase superconducting synchronous/induction motor with a liquid helium cooled rotor and a liquid nitrogen cooled stator. An ordinary induction motor equivalent circuit was used to describe starting before its transition to superconductivity. They concluded: 1) That the motor will not operate if the rotor is superconducting at starting, that it needs rotor resistance to start. 2) That the rotor and stator must have iron flux paths, unlike superconducting synchronous machine designs. 3) That the motor will operate similarly to a synchronous machine after the rotor conductors have become superconducting. They did not attempt to describe the period during the transition to superconductivity. The latter part of the article considered the thermal design of the rotor. No operating models or prototypes were mentioned.

In 1987 De Doncker and Novotny[6] wrote a paper describing power applications of high T_c superconductors. In a section on induction machines they made the following points about typical superconducting synchronous/induction motor designs: 1) Once induction motor rotor windings become superconducting, no flux changes can occur in the rotor windings. 2) Superconducting synchronous/induction machines lose superconductivity in their rotors at starting due to excessive rotor currents but become superconductive at low slips when rotor currents have dropped. 3) Superconducting operation is similar to that of a synchronous machine with a fixed field. 4) In usual designs stator windings are cooled to liquid nitrogen temperatures but are not superconducting.

In 1987 Lipo[7] discussed the possibility of a high T_c superconducting synchronous/induction motor. He considered the question of rotor stability after the transition of a superconducting motor to superconductivity. He produced graphs that showed significant torque and speed oscillations during the starting of an ordinary induction motor and during the starting of a proposed high T_c superconducting synchronous/induction motor.

1.7.2 High T_c Superconducting Motors

Jordan[8] has compared the feasibilities of different types of high T_c superconducting machines. He concludes that synchronous motors are the most feasible. His company, the Reliance Electric Company, has begun a feasibility study of them. Jordan also concludes that a high T_c superconducting wire must be developed to make these synchronous motors possible.

A type of universal motor using the new high T_c superconductors was built by Poeppel of the Argonne National Laboratory. According to Poeppel, "It's too small for practical use and produces negligible power, but it demonstrates for the first time that these motors are possible." It operated similarly to a universal motor with fields produced by armature coils repelling fixed superconductor "poles" on the stator by the Messner effect. The most descriptive paper found on the motor was written by Nailen[9].

Although the motor was of a different type than the superconducting synchronous/induction motor considered in this dissertation some conclusions can be extrapolated from Poeppel's motor. First, induction motors using the new high T_c superconductors should be possible. Second, an induction motor built with the present lower capacity high T_c superconductor material will also produce a negligible power.

1.7.3 Superconducting Generators

There has been a great deal of research done on liquid helium cooled superconducting generators. Multimillion dollar research programs have been carried out by large corporations in many countries. Much of this research was done in the 1970's. Good coverage of superconducting generators can be found in the books by Bumby[10] and Nasar[11].

Although the previously researched superconducting generators have some obvious similarities to the superconducting synchronous/induction motors of this dissertation there are significant differences. First and most important, the superconducting generators did not have "trapped flux" as the superconducting synchronous/induction motors do. Second, the rotors of the superconducting generators had an electrical connection to a normally conducting external DC source. This is different from the superconducting synchronous/induction motor rotors that have no connection to an external source except for the magnetic links they have with the stator. Third, the previously researched superconducting generators operated with liquid helium cooled rotors. That introduced thermal difficulties that do not have to be considered in the warmer liquid nitrogen temperature superconducting synchronous/induction rotors of this dissertation. Fourth, the magnetic flux densities used on the superconducting generators were much higher than those that can be used with the iron structure of the superconducting synchronous/induction motor. Fifth, the air core design of the superconducting generators and the high fields used required the use of a containment ring of magnetic iron around the generator structure. No containment ring is necessary around the superconducting synchronous/induction motor, since it is just like a standard induction motor in that regard.

Because of the above differences detailed library research was not done on superconducting generators.

1.7.4 High Speed Motors

High speed in this dissertation means above 9,000 rpm. No mention was found of any work on superconducting high speed motors prior to this dissertation. All of the high speed motor papers reviewed considered room temperature, normal conductor AC motors.

1.7.4.1 High Speed Induction Motors. In 1971 Hitachi[12] announced that it had produced a 50 W solid rotor AC motor that could attain speeds of 500,000 rpm. It used air bearings. This was the fastest motor found during the literature search. For this dissertation 500,000 rpm serves as a practical upper speed limit for a small motor.

In 1974 Lakerveld[13] described the high speed solid rotor induction motors he built. His motors developed maximum speeds of 40,000 rpm from a 700 Hz square wave two phase supply. Lakerveld observed that rotor loss inducing harmonics are produced by the unsymmetrical structure of stator teeth. In his motors he reduced rotor losses by making his two stators without conventional rotor teeth. One used thin flat coils laid into the air gap and glued to the stator iron. The other used toroidal coils and large pole faces on the stator.

Hesmondhalgh and Tippings[14,15,16] have done some significant work on high speed induction motors.

Hesmondhalgh and Tippings's first paper[14], published in 1983, described theoretical and experimental research they did on miniature induction motors capable of maximum speeds of 150,000 rpm. They showed theoretically that a sheet rotor (rotor with a conductor sheath around it instead of rotor bars) is inferior to the squirrel cage rotor. However, they decided that the extra complications of building a squirrel cage rotor did not warrant the use of squirrel cage rotors on their miniature machines and used the simpler sheet rotors. Although they used toothed stators they mention the tooth structure as a problem that induces magnetic losses. A major problem mentioned was the bearings. They wanted to use air bearings but had mechanical resonance problems and instead decided on oil impregnated paper based phenolic comb bearings. These bearings were said to be good to 500,000 rpm. The only problem with the phenolic comb bearings was their extremely small fit tolerance. They wrote that micron gap interference fits were necessary. They wanted to produce an output of 10 W at 150,000 rpm but have concluded that without resorting to forced cooling or finding some way to reduce rotor magnetic losses, it can not be done with an induction motor.

Hesmondhalgh and Tippings' second paper[15] repeats the observations and conclusions of their first paper[14] and adds more. The most significant addition is the description of the transformer coupled stator. To produce the highest possible slot packing factor and thereby reduce copper losses to a minimum transformers were used to send high currents into single conductor stator coils.

A number of researchers have reported on more powerful higher speed induction motors. A much more powerful motor was mentioned by Tanaka and Kuga[17]. In 1985 they reported on a 2000 kw 8000 rpm squirrel cage induction motor built for Toshiba. Solid rotor induction motors were mentioned by Petit and Delassus[18] in a 1973 paper. Petit and Delassus reported on motors built for CEM that were rated for 147 kw,

12,000 rpm and 100 kw, 30,000 rpm. Solid and laminated rotor induction motors were built and compared in a 1990 paper by Ikeda, Sakabe, and Higashi[19]. Their motors were rated for up to 200 kw output at 12,000 rpm. Ikeda, Sakabe, and Higashi concluded that the laminated rotor is superior to the solid rotor in this application.

1.7.4.2 <u>High Speed Reluctance Motors</u>. In 1989 Fukao, Chiba, and Matsui[20] published the results of tests they did on their high speed reluctance motors. They found that by using amorphous-iron in the stator and rotor they were able to reduce the core losses to one fifth of what they are in a comparable high speed ordinary silicon steel reluctance motor. Their test motor was capable of 48,000 rpm with an output of 372 W at an efficiency of 85%.

Their work demonstrates the advantage of amorphous-iron for reducing high frequency core losses.

1.7.4.3 <u>High Speed Synchronous Motors</u>. Hesmondhalgh and Tippings[16], after concluding that they could not make a 150,000 rpm, 10 W, induction motor designed and built a similarly sized 150,000 rpm 100 W permanent magnet synchronous motor. They used the same transformer coupled stator design as before. The two pole rotor used a samarium cobalt magnet. The use of the high flux density permanent magnet rotor allowed the stator supply current to be decreased. Reduced stator current then reduced the stator resistive power losses. Also the synchronous operation of the permanent magnet rotor eliminated rotor losses.

Higher power high speed permanent magnet motors have been built. Pouillange[21] mentions a 48,000 rpm, 100 kW permanent magnet motor. Matsuda and Mizuno[22] mention a 40,000 rpm, 1 kW permanent magnet motor.

1.7.4.4 High Speed Motor Power Limitations. Matsuda and Mizuno[22] presented an empirical equation that determines an approximate maximum motor power limit for a given motor speed. The equation is:

$$P_{Mout} = \frac{K_M}{W_M^{2.8}} \qquad (1\text{-}1)$$

K_M = maximum speed-power constant

P_{Mout} = maximum speed-power motor power capacity (kw)

W_M = maximum speed-power motor speed of revolution (rpm)

Using a per unit base of P_{Mout} = 10 kw and W_M = 40,000 rpm, K_M was stated to be 0.4 for permanent magnet motors, 1.0 for induction motors and 3.0 for reluctance motors.

The equation was tested against induction motors built by others. Figure 2 graphs induction motor power capacity versus operating speed. It contains a curve produced by Matsuda's equation and the experimental results of four other authors. As can be seen Matsuda's equation underestimates motor capacities that some other researcher's motors produced. However, it is still felt that the equation can be used for determining reasonable initial estimates.

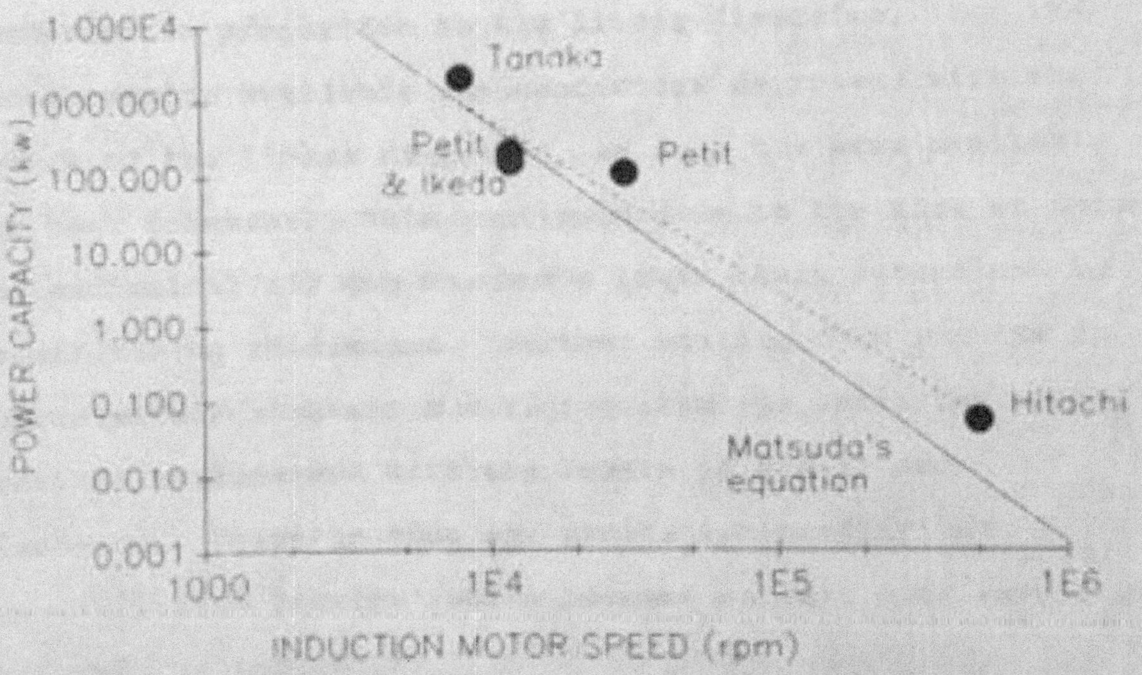

Figure 2 Curve of Matsuda's equation for power capacity versus induction motor speed. Points are from Hitachi[12], Ikeda[19], Petit[18], and Tanaka[17].

1.7.5 Power Limitations of Small Induction Motors

There are limits as to how far an induction motor design can be scaled down. These limits apply to the prototype motors built for this dissertation.

They were best explained by Miller[23]. Quoting Miller: "The efficiency and power factor of small motors falls off in small sizes because of the natural laws of scaling, particularly at part load. As a motor of given geometry is scaled down, if all dimensions are scaled at the same rate the m.m.f. required to produce a given flux-density decreases in proportion to the linear dimension. But the cross-section available for conductors decreases with the square of the linear dimension, as does the area available for heat transfer. This continues down to the size at which the mechanical air gap reaches a lower limit determined by manufacturing tolerances. Further scaling-down results in approximately constant m.m.f. requirements while the areas continue to decrease with the square of the linear dimension. There is thus an 'excitation penalty' or 'magnetization penalty' which becomes rapidly more severe as the scale is reduced."

1.7.6 Normal Conducting Cryogenic Motors

A nearly commercial motor has already been built to operate at near liquid nitrogen temperatures. Yoda, Kamata, and Shimizu[24] reported on a totally immersed liquified natural gas pump motor that they built and tested in operating conditions for 8,000 hrs. The motor was a 3,000 volt, 2 pole, 50 Hz, 150 kW, 3 phase induction motor. Since the temperature of liquified natural gas is -162° C, only 34° C above that of liquid nitrogen it is logical to assume that a large liquid nitrogen temperature motor can be built as well. Notable features of their motor were: 1) Special hydrostatic ball bearings that used liquified natural gas as their lubricant. 2) Formed stator coils insulated with mica tape/epoxy insulation.

1.7.7 Characteristics of Superconductors

The zone of a material's superconducting state is bounded by: temperature, magnetic flux density, and current density. These have been plotted for various superconducting materials in three dimensional graphs such as those presented by Fitzgerald[25].

The zone of a material's superconducting state is also bounded by the frequency of the magnetic flux and current applied. Superconductors experience hysteresis power losses similar to those in seen in magnetic materials. Hysteresis losses are proportional to frequency. If the hysteresis losses are too high a superconductor will overheat and be driven out of superconductivity.

Frequency sensitivity has been noted by many researchers investigating low T_c superconductors. In 1975 in reference to low T_c superconductors, Rechowicz[26] wrote

"Even with considerable refinements such as the introduction of high-resistivity cupro-nickel around the filaments to help reduce the electric coupling between individual filaments, the high losses prevent the use of filamentary superconductors at frequencies greater than 1 Hz." More recent developments with low T_c superconductor materials now allow significant 60 Hz currents to be carried with special cables made of improved twisted and transposed bundles of thin filaments of low T_c superconductor material. The thin filaments are on the order of 9 μm in diameter. Some recent papers on low T_c superconductor AC cables are by Mower and Iwasa[27], and Yamamoto et al.[28].

Hysteresis and its corresponding frequency sensitivity occur in high T_c superconductors. The existence of hysteresis in high T_c superconductors has been surmised from AC susceptibility measurements made by Abell et al.[29], Goldfarb et al.[30] and others. Cave et al.[31] have used low frequency magnetization curves they measured on sintered $YBa_2Cu_3O_x$ to estimate AC losses at 50 Hz. They predict AC losses in $YBa_2Cu_3O_x$ experiencing .2 T peak at 50 Hz to be 100 times greater than those in ultrafine NbTi superconductors under the same conditions. Cave et al.[31] also predict that for $YBa_2Cu_3O_x$ to be useful in AC applications it should be made into cables with filaments of less than 10 μm in diameter.

1.7.8 General Discussions on High T_c Superconductors and Their Applications

Fisher[32] has written a paper that brings the reader to date on the current state of the art of high T_c superconductors in April, 1988. An extensive bibliography is included.

Schlabach[33] has written a paper high T_c superconductor wire and large scale applications of high T_c superconductors.

Two articles one in the Spectrum magazine, Fitzgerald[25], and one in Power Engineering, Smock[34], discuss the whole field of high T_c superconductors generally. An article by Sabrie[35] discusses the state of the art of the application of superconductors to machinery in March 1987.

2.0 DESIGN REQUIREMENTS AND DECISIONS

2.1 Rotor Resistance and Inductance

For a superconducting synchronous/induction motor to start as an induction motor it must have the ferromagnetic rotor structure of an induction motor. This must be to help achieve rotor resistance and inductance for induction starting. However, the use of ferromagnetic materials also introduces the possibility of saturation. A superconducting rotor using standard laminations has a maximum flux density of about 2 T. This helps place a limit on the output power capacity of a superconducting synchronous/induction motor since power is proportional to flux.

A superconducting synchronous/induction motor also needs rotor resistance to start as an induction motor. In this research the material of the superconducting portion of the squirrel cage is kept out of superconduction during initial starting by not cooling it.

2.2 Rotor Structure

In this research the structures of the rotors are the same as those of existing induction motor rotors. The same teeth and gap shapes and dimensions of existing induction motors are used. The reason for this is to allow the use of existing induction motor models with less probability of unforseen errors.

When actual superconducting motors are built it may be desirable to change the ratio of the rotor bar material to the lamination material, but this will not be considered here.

2.3 Rotor Speed and Power

The faster a rotor revolves the more power it can develop. However, a high speed rotor will experience greater centrifugal forces than a conventional lower speed rotor of the same size. The greater forces will tend to pull the rotor bars out and if the rotor is not perfectly balanced, tend to overload and destroy the bearings.

In an actual rotor there are also more complicated forces. If a rotor is balanced, but not balanced uniformly along its length significant centrifugal forces may still be experienced. If the distance between the bearings is long relative to the rotor diameter a resonant vibration may occur. With this the rotor may vibrate in a wave like manner similar to a guitar string. This may also be initiated by the interaction of the rotor and its bearings, a condition similar to this was mentioned by Hesmondhalgh and Tippings[14]. Resonant vibrations could also be initiated by slightly imbalanced magnetic forces from a stator that has slight differences in its winding geometry or magnetic path reluctances.

All of the imbalance problems can be reduced by better machining, better assembly, and higher quality parts. However, decreasing imbalance this way increases expense and produces a less robust motor. To keep the high speed designs realistic, operating speeds of the high speed motor designs of this dissertation were set by empirical methods.

2.4 Rotor Cooling System

Rotor superconducting materials must be kept below their critical temperature during steady state operation. For the new high T_c superconductors assumed in this dissertation liquid nitrogen temperatures are sufficient. The designs of the motors in this dissertation use temperature to switch the superconductor materials into superconductivity. A controllable cooling system is specified.

The most cooling is needed during starting when the rotor bars are not superconducting. At this time rotor conduction and magnetic losses are taking place. After starting conduction and most magnetic losses in the rotor will drop to zero. Then the heat received by the rotor will be from windage, rotor surface magnetic heating, and heat conveyed across the air gap from the stator. The necessary closeness of the rotor laminations to the stator laminations means a substantial amount of heat may be conducted to the rotor from the stator. Some attention must be paid to the stator design to reduce the amount of waste heat it sends to the rotor. Using thin, large, low magnetic hysteresis loss stator laminations, and liquid nitrogen cooled low resistance stator windings is called for.

In liquid helium cooled generators, liquid helium was pumped into and out of the turning rotor. This required close tolerance fits between the rotating rotor and the stationary liquid helium supply. This way the rotor was more assured to be cold and the expensive liquid helium coolant was not wasted in cooling the stator coils. However, the difficulty of making a coolant seal good enough to contain liquid helium between a rotating and a stationary system, insulating a turning rotor from the stator and having a coolant pumping system are substantial[36]. The problems of handling liquid nitrogen coolant are not as serious in the liquid nitrogen cooled high T_c superconducting synchronous/induction motors of this dissertation. Liquid nitrogen is less expensive and warmer.

A ratio is calculated for the effective power required to create liquid nitrogen to that expended in cooling the rotor. This is done with a commercial electric energy rate, a commercial cost of liquid nitrogen per mass and the vaporization energy per mass of liquid nitrogen. The equation is:

$$R_{LN/CV} = \frac{R_{LN}}{C_N R_P} \qquad (2-1)$$

R_P = commercial electricity rate from Duquesne Light Co., Pittsburgh, PA $2.661 \times 10^{-8} \frac{\$}{\text{Joules}}$

C_N = heat of vaporization of liquid nitrogen at 77° K[37] $1.98 \times 10^{5} \frac{\text{Joules}}{\text{kilogram}}$

R_{LN} = cost of liquid nitrogen, cost from Air Products, Inc., Pittsburgh, PA .457 $\frac{\$}{\text{kilogram}}$

$$R_{LN/CV} = \frac{.457}{(1.98 \times 10^{5})(2.661 \times 10^{-8})} = 86.74$$

The power ratio calculated in this way might appear overly conservative. It could be argued that a liquid nitrogen supply system close to the motor could be used. The superconducting synchronous/induction motor itself might be used to power the nitrogen compressor. However, even with an ideal compressor the ratio will be greater than 1 and an actual compressor would produce a ratio far greater than that. Furthermore, the cost of purchasing and maintaining a liquid nitrogen system would have to be included. Therefore it is felt that using commercial electric energy and liquid nitrogen rates produces a realistic indication of the actual effective power drain of the cooling system.

The magnitude to the ratio indicates the importance of keeping the waste heat going to the rotor's superconductors to a minimum. The superconducting portion of the rotor should be as thermally insulated as possible.

2.5 Rotor Bar Superconductor Qualities

A superconductor must have a number of qualities that it can maintain for many years to be acceptable as rotor bar material.

2.5.1 Temperature, Magnetic Flux Density, Current Density, and Frequency Bounds

A superconductor must remain superconductive within maximum temperatures, magnetic flux densities, current densities, and frequencies that will exist during motor operation. Some values for these will be determined in chapter 5.0.

2.5.2 Mechanical, Thermal, and Environmental Ruggedness

The superconductor must be mechanically rugged. It must be able to survive the inevitable and continuous vibrations it will experience during motor operation and the sudden impact forces it will receive during starting.

The superconductor must be thermally rugged. During steady state operation it will be at a constant temperature near 77° K. However, it will have to endure thermal cycling with starting and stopping and it will probably be stored at room temperature. Changing temperatures will cause differential expansion between outer and inner areas of the superconductor and between the superconductor and the material it rests in.

Present high T_c superconductors are degraded and eventually destroyed by water. If the present materials are used, they must be protected. Only dry air can be allowed into the presence of the superconductor materials or the materials must be covered with a waterproof coating.

3.0 MOTOR CONSTRUCTION MATERIALS

Whenever possible standard construction materials were specified for the motors of this dissertation.

3.1 Available Superconductor Materials

High T_c superconductor materials are a recent development. The first one, $YBa_2Cu_3O_x$, was created in February of 1987. Now there are other types available that are approximately equal in their characteristics. Among these are $BiSr_2CaCu_2O_x$, $BiSrCaCu_2O_x$, $Bi_2Sr_2CaCu_2O_x$, $BiSrCaCu_2O_x$, $Tl_2CaBa_2Cu_2O_x$. All are composed of rare earth elements combined with copper and oxygen into a crystalline structure. They are type II superconductors. Their superconducting form is ceramic. Bulk produced forms of these materials resemble graphite in appearance. The transition temperature for these materials is between 92 and 115° K. Improvements are being made on these materials and more are still being developed.

For use in motor rotors all of the high T_c superconductors present similar problems.

The first problem is that superconductors are sensitive to the frequency of applied currents and magnetic fields. Superconductive rotor bars in a superconducting squirrel cage synchronous/induction motor will have to carry AC, even during steady state operation.

Magnetic hysteresis heats superconductors and tends to drive them out of superconductivity. All superconductors, low and high T_c, are sensitive to AC. Cables formed from twisted and transposed superconductor filaments can withstand higher frequencies than solid superconductor material. However, superconductors formed into cables of filaments can not be used in squirrel cages, since there is no way to connect the filaments into superconducting junctions.

The second problem is superconductor materials have critical (maximum) current and magnetic flux densities. These values are almost always quoted for DC currents and magnetic fields. The following values, reported by Fitzgerald[25] on work done at AT&T Bell Laboratories, apply to ceramic superconductor materials at 77°K. At Bell, thin film superconductors have been made with maximum current densities up to 4×10^{10} A/m^2. The larger bulk produced superconductors, that are big enough for producing parts like motor squirrel cages have lower maximum current densities. The melt-textured bulk superconductors they made have maximum current densities up to 1.7×10^8 A/m^2 in zero magnetic field. The maximum current density that could be applied to their bulk sintered superconductor material was on the order of 1.6×10^7 A/m^2. Magnetic fields applied to a superconductor decrease its maximum current density. The best ceramic superconductor they tested, the thin film type, remained superconducting in fluxes up to 10 T.

Fisher[32], in a review of the works of many researchers, reported lower values. He reported maximum current densities on the order of 2.5 x 10^6 A/m^2 for sintered bulk superconductors. The highest current densities he reported were for powdered superconductors, on the order of 5 x 10^8 A/m^2. He reported a low maximum magnetic flux density for bulk sintered superconductors of less than 1 x 10^{-4} T. He mentioned, it is believed that the same bulk sintered material has grains of better superconductor in it and that these will remain superconductive in flux densities greater than 10 T. This dissertation author's experience with testing commercially available large bulk sintered superconductors indicates the lower values are presently more applicable.

The third problem is that special care must be used in shaping the superconductors. So far, a method of making superconductor to superconductor superconducting joints has not been developed. Furthermore the material may undergo considerable shrinkage during processing. For example, after being pressed into a solid form a bulk type superconductor may shrink 20% during sintering. This means that the superconductor material must be formed in a large block and machine ground into shape. The high T_c superconductor, like a ceramic, requires grinding rather than cutting as the machining method.

The construction of a standard 65 bar squirrel cage rotor would be very tedious, the superconductor materials would have to be machined to a squirrel cage shape. This is much more difficult than the usual processes of casting or soldering done with ordinary induction motor aluminum, copper or brass squirrel cages.

Fourth, high T_c superconductor materials are damaged by water. If they get wet they lose their superconductivity ability. Care must be taken during machining and use. Probably a waterproof coating should be applied to a superconductor before it is used.

Fifth, high T_c superconductor materials have a tendency to crack as they age. Once cracked they no longer superconduct.

Sixth, high T_c superconductor materials are not mechanically strong, care would have to be taken to avoid its cracking or breaking while being machined and used in a rotating rotor.

3.2 Normal Conductor Material Characteristics versus Temperature

No difficulties were found with the copper and lead/tin solder conductor materials used the prototype stators and rotors while they were used at liquid nitrogen temperatures. Also no difficulties were mentioned in the work of Yoda, Kamata, and Shimizu[24] on their liquified natural gas pump motor.

There is a resistance reduction advantage to using normal conductor materials at liquid nitrogen temperature. For example, the resistance of copper at that temperature drops to about 1/6 of its room temperature resistance. The 1/6th resistance drop reduces resistive power losses to 1/6th of their room temperature value.

3.3 Magnetic Materials

3.3.1 Characteristics versus Temperature

Liquid nitrogen temperatures have no significant effect on the magnetic properties of the stator and rotor magnetic materials. None were seen in this research or mentioned in the paper of Yoda, Kamata, and Shimizu[24] on their liquified natural gas pump motor.

3.3.2 Characteristics versus Frequency

High speed motor operation requires high frequency reversals of magnetic flux through the magnetic materials. This increases the eddy current and hysteresis losses. Two material specifications help minimize these losses. The use of thinner laminations in the stator and rotor will reduce eddy current losses. The use of magnetic laminations with the smallest hysteresis loops will reduce hysteresis losses. Hysteresis loops are reduced in size by selecting the right material and using more material per flux.

A stator and rotor made of amorphous-iron such as used be Fukao, Chiba, and Matsui[20] would reduce magnetic losses at higher frequencies.

3.4 Electrical Insulation Characteristics versus Temperature

Most electrical insulation retains its insulating value at liquid nitrogen temperatures. However, insulation rated for room temperatures often does not retain its mechanical strength and adhesive properties at liquid nitrogen temperatures.

It was observed that room temperature red insulating enamel that was painted onto wires quickly fell off when cooled to liquid nitrogen temperatures. Also, epoxy rated for room temperature use fell off materials it was previously stuck to when cooled to liquid nitrogen temperatures.

Special insulating materials that remain strong, flexible, and adhesive at low temperatures are needed. Two insulating materials were found to work well at liquid nitrogen temperatures by this author. The thin layer of insulation on standard heavy armored polythermalene magnet wire remained intact and flexible when the wire was immersed in liquid nitrogen. This wire was used in the prototype stators. A special epoxy manufactured by the Emerson & Cuming Company called "Stycast Blue" was found to retain its mechanical strength and adhesive qualities at liquid nitrogen temperatures. It was used to pot the stator windings and hold the stator stack in place in the prototype motors. Also, Yoda, Kamata, and Shimizu[24] mentioned the successful use of a mica tape/epoxy insulation in their liquified natural gas pump motor.

3.5 Bearings

The selection of the type of bearing depends on the motor design and application.

3.5.1 Characteristics versus Temperature

In the prototype motors the bearings were cooled to liquid nitrogen temperatures. Motors where the bearings will be cold can not have the usual oil or grease lubricant in their bearings. Bearings with dry graphite lubricant or with another special lubricant can be used. The ball bearings of the low temperature prototype motors used a teflon based dry lubricant that was coated on the ball retainer ring. Yoda, Kamata, and Shimizu[24] on their submerged liquified natural gas pump motor used ball bearings submerged in and lubricated by liquified natural gas.

It is also possible to design superconducting synchronous/induction motors with the standard room temperature bearings that are insulated from the cold rotor and stator. Liquid helium temperature superconducting generators were built this way.

3.5.2 Characteristics versus Speed

Bearings have maximum rpm ratings. They are a major limitation in the construction of high speed motors. Radial and axial load ratings of bearings are approximately inversely proportional to their maximum rpm ratings. The highest speed bearings use air or magnetic levitation, but these bearings are not very robust, they require close tolerances in construction and can not handle loading well.

4.0 INDUCTION MOTOR WITH A NORMAL CONDUCTING ROTOR MODELLED WITH A FIXED SINUSOIDAL ROTATING FIELD TO ESTIMATE SUPERCONDUCTOR REQUIREMENTS

Before a superconducting synchronous/induction motor makes the transition to superconductivity it operates as an ordinary induction motor. This chapter considers a superconducting motor in the ordinary induction motor mode as it is in steady state induction motor operation between full load and no load.

The analysis method considers one motor slip at a time. At each slip an initial constant magnitude sinusoidal air gap rotating flux wave is estimated. That flux wave is used to calculate induced voltages on both stator and rotor conductors. Finally the magnitude of the flux wave is adjusted iteratively until a proper stator input voltage is calculated.

Magnetic flux densities and current densities experienced by the rotor bars are calculated. The calculated values are the maximum that a superconductor material would have to withstand to go into superconductivity from normal conductivity. A later chapter, 5.0, considers values that are the maximum that a superconductor material would have to be able to withstand to stay superconductive.

Starting from zero speed is not considered, since the motor becomes superconducting at near synchronous speeds.

Transient conditions are not considered here. The transient operation of the transition to superconductivity is considered in chapter 7.0.

The rotor bars here are assumed to be made of normal and superconductive material. As indicated in Figure 1 the outer 1/6th of each rotor bar and end-ring is copper, the inner 5/6th is superconductor material. The composite squirrel cage is treated as an ordinary normal conductor squirrel cage. Here the effective squirrel cage always has a finite resistance, even if parts of it go into superconductivity. This analysis could be applied to an ordinary induction motor.

4.1 Approximations and Parameters

A number of simplifying approximations are made in the analysis of this motor. Unless otherwise noted the same approximations are made in this chapter as in 5.0 and 6.0.

Only the fundamental component of the flux density wave is considered.

This method of analysis uses the approximation that the number of rotor bars are integer multiples of the number of stator poles and number of phases. This simplifies calculations by allowing the rotor to be considered as having symmetrically placed groups of perfect pitch windings. In reality induction motors are not built with these integer multiple numbers of rotor bars. Reluctance torque could keep them from self-starting and cause cusps in their speed torque curves if they did rotate. These torque effects have been reported by many authors, among these are Alger[38], Hamilton[39], and Kuhlmann[40]. However, using the approximation of perfect pitch rotor coils greatly simplifies the analysis. An actual rotor does not have an integer number of perfect pitch coils. In the calculations the assumption creates fractional numbers of perfect pitch rotor coils.

Rotor bars in this chapter never form circuits that are superconducting throughout, to do so would trap flux. Rotor bars that form complete superconducting circuits are considered in chapter 5.0 and 6.0. However, during analysis the performance of the model is considered as the effective rotor bar resistance is reduced. Resistance reduction could happen as sections, but not continuous loops, of the squirrel cage become superconductive.

It is assumed that the cooling system keeps the rotor near 100° K in this chapter.

Stator and rotor resistances and leakage impedances are assumed constant in this analysis.

4.2 Equations

Superconducting synchronous/induction motors make a

transition from normal conduction to superconduction operation near synchronous speed. Analysis of rotor bar flux densities and current densities of this motor are done in that range.

An air gap flux density wave rotates around at synchronous speed. It produces currents in the rotor bars. The currents in the rotor bars interact with the rotating flux density wave and push the rotor around. The same air gap flux density wave produces a voltage in the stator circuit. The stator produced voltage helps determine the stator input current.

The equations have been written for a 2 pole motor. However, they can be applied to motors with more poles by applying appropriate multiplication factors.

The equations of 4-1 to 4-31 have been entered into an iterative FORTRAN program to solve for motor quantities.

All equations use peak values unless otherwise stated.

4.2.1 Approximate Total Flux

The upper limit of the total peak flux passing through a rotor is:

$$\phi_M = \frac{2E_T}{k_d k_p N w_s} \tag{4-1}$$

It will be used as an initial estimate for ϕ_M.

4.2.2 Rotor Voltage Generated per Effective Rotor Coil

The rotor is assumed to be made of a collection of symmetrically placed, perfect pitch, single turn coils. With that assumption the voltage generated in each coil is:

$$E_r = \phi_M W_s s \tag{4-2}$$

4.2.3 Stator Voltage Generated per Phase

The back emf voltage generated by the rotating flux on each stator winding is:

$$E_{Tg} = .5 k_d k_p N W_s \phi_M \tag{4-3}$$

4.2.4 Rotor Current per Effective Rotor Coil

The current in each rotor coil can be calculated with the generated rotor voltage and the rotor coil's effective impedance. The rotor coil's impedance is composed of resistance and leakage inductance. In the range from no load to full load it is reasonable to assume the resistance of each rotor coil is constant near its DC value, since the frequency seen by the rotor coils is low. The leakage inductance of the rotor coils depends on the shape and dimensions of the flux paths and the degree of saturation of the stator and rotor lamination materials. Generally the leakage inductance of the rotor circuit is assumed constant in the operating range. The assumption of constant leakage inductance in the operating range is used here.

Rotor bar resistance is calculated from its rotor material resistivity and dimensions. Rotor bar leakage inductance is calculated from rotor dimensions, assumed flux density and empirical equations. The values of end-ring resistance and inductance are included with those for the rotor bars.

Applying the induced rotor voltage to a rotor coil the current is:

$$I_{rb} = \frac{E_r}{2[R_{rb}^2 + (L_{rb}w_s s)^2]^{.5}} \qquad (4\text{-}4)$$

4.2.5 Electrical Torque Created by the Interaction of Flux and Rotor Current

The electrical torque is the torque produced by the interaction of the current through each rotor bar and the flux density it is in. It is does not consider the torque lost by friction. On a three phase machine it can be determined by solving for the torques produced by the sums of symmetrically spaced rotor coils and then summing those torques for all of the rotor coils.

Instantaneous currents through 3 symmetrically placed groups are as follows:

$$i_{ab}(t) = \frac{w_s s \phi_M}{2[R_{rb}^2 + (w_s s L_{rb})^2]^{.5}} \cos\left[w_s st - \operatorname{atan}\left(\frac{w_s s L_{rb}}{R_{rb}}\right)\right] \qquad (4\text{-}5)$$

$$i_{bb}(t) = \frac{w_s s\phi_M}{2[R_{rb}^2 + (w_s sL_{rb})^2]^{.5}} \cos\left[w_s st - \operatorname{atan}\left(\frac{w_s sL_{rb}}{R_{rb}} + \frac{2\pi}{3}\right)\right] \quad (4\text{-}6)$$

$$i_{cb}(t) = \frac{w_s s\phi_M}{2[R_{rb}^2 + (w_s sL_{rb})^2]^{.5}} \cos\left[w_s st - \operatorname{atan}\left(\frac{w_s sL_{rb}}{R_{rb}} - \frac{2\pi}{3}\right)\right] \quad (4\text{-}7)$$

Instantaneous flux densities experienced by rotor bars of the same 3 symmetrically placed groups are as follows:

$$B_{ab}(t) = \frac{\phi_M}{2rl_g} \cos(w_s st) \quad (4\text{-}8)$$

$$B_{bb}(t) = \frac{\phi_M}{2rl_g} \cos\left(w_s st + \frac{2\pi}{3}\right) \quad (4\text{-}9)$$

$$B_{cb}(t) = \frac{\phi_M}{2rl_g} \cos\left(w_s st - \frac{2\pi}{3}\right) \quad (4\text{-}10)$$

The torque produced by this group can be calculated as:

$$T_g = 2rl_g[B_{ab}(t)i_{ab}(t) + B_{bb}(t)i_{bb}(t) + B_{cb}(t)i_{cb}(t)] \quad (4\text{-}11)$$

The above equation produces a constant torque for all times. It reduces to:

$$T_g = \frac{.75 w_s sR_{rb}\phi_M^2}{[R_{rb}^2 + (w_s sL_{rb})^2]^{.5}} \quad (4\text{-}12)$$

There are a number of symmetrical groups each producing

a constant T_g. Multiplying T_g by the number of rotor bars per phase divided by 2 produces the total electrical torque.

$$T_e = \frac{.375 N_b p w_s s R_{rb} \phi_M^2}{m[R_{rb}^2 + (w_s s L_{rb})^2]^{.5}} \tag{4-13}$$

4.2.6 Power Produced by the Electrical Torque

The effective power produced by the electrical torque is:

$$P_e = T_e w_s (1 - s) \tag{4-14}$$

4.2.7 Power Going from the Stator to the Rotor and to Turn the Rotor

All power used in the rotor and to turn the rotor comes from the stator. It is:

$$P_{rt} = P_e + \frac{N_b p}{2} I_{rb}^2 R_{rb} \tag{4-15}$$

4.2.8 Stator Magnetizing Current

An equation expressing magnetizing current in terms of total flux per pole was composed of equations from Kuhlmann[41]. His method was to sum the mmf drops of each part of the magnetic circuit. Drops were found for the lamination paths in the rotor and stator and in the air gap. As he showed for a motor at full load, most of the mmf drop is in the air gap. His example motor had about 80% of the magnetizing current caused by the air gap. The magnetizing current equation used here assumes all magnetizing current is caused by the air gap. It is felt that the 20% error in the magnetizing current is acceptable since it results in only an approximate 5% error in the resultant stator current. A correction factor is applied to the equation when it is used on the example motor to raise the magnetizing current to the magnetizing current value predicted for the motor at full load. The equation for magnetizing current is:

$$I_M = \frac{8.819 \times 10^5 \phi_M \delta k_r k_s p^2}{m_s N k_d k_p D l_g} C_{orf} \qquad (4\text{-}16)$$

4.2.9 Stator Current

The stator current is composed of the magnetizing current and a reflected value of the rotor current.

$$I = \left(\left\{ I_M + \frac{I_{rb} N_b p}{m N K_d K_p} \frac{L_{rb} w_s s}{[R_{rb}^2 + (L_{rb} w_s s)^2]^{.5}} \right\}^2 + \left\{ \frac{I_{rb} N_b p}{m N K_d K_p} \frac{R_{rb}}{[R_{rb}^2 + (L_{rb} w_s s)^2]^{.5}} \right\}^2 \right)^{.5} \qquad (4\text{-}17)$$

4.2.10 Angle Between the Stator Current and the Stator Voltage Generated at the Gap

The voltage generated on the stator is not in phase with the current through it. The angle can be calculated from an effective power factor by dividing real power from the stator by va's from the stator. The angle is:

$$A_{sgvi} = -\text{acos}\left(\frac{P_{rt}}{1.5E_{Tg}I}\right) \tag{4-18}$$

4.2.11 Calculated Stator Input Voltage

The input voltage to the stator is calculated with the phasor equation:

$$\mathbf{E}_T = \mathbf{E}_{Tg} + \mathbf{I}(R_s + w_s L_s j) \tag{4-19}$$

Where phasor values are:

$$\mathbf{E}_{tg} = E_{tg}\angle 0 \tag{4-20}$$

$$\mathbf{I} = I\angle A_{sgvi} \tag{4-21}$$

$$\mathbf{E}_T = E_T\angle \delta_p \tag{4-22}$$

4.2.12 Mechanical Output Power

The mechanical output power is the electrical power minus the friction losses. Friction losses are proportional to speed. An empirical equation is used to approximate that effect as slip decreases. The equation is:

$$P_{out} = P_e - P_{fws}(1 - s)^{1.35} - P_{fl} \tag{4-23}$$

4.2.13 Mechanical Output Torque

The output torque is determined by the relationship:

$$T_{out} = \frac{P_{out}}{(1 - s)w_s} \tag{4-24}$$

4.2.14 Stator Lamination Magnetic Losses

During normal operation the frequency of the magnetic flux in the rotor is much lower than in the stator. Since magnetic losses are proportional to frequency the losses are much smaller in the rotor, even though the rotor has similar flux densities. For that reason only the magnetic losses in the stator need to be considered.

The value of the average rms flux density in the stator depends on the dimensions of the stator and the flux it carries. Solving for this requires a magnetic circuit analysis that is particular to the machine being analyzed. Details on the procedure are described in section 4.3 for the example motor.

Once the average rms flux density is known it can be applied to a curve such as the one given in Figure 3.

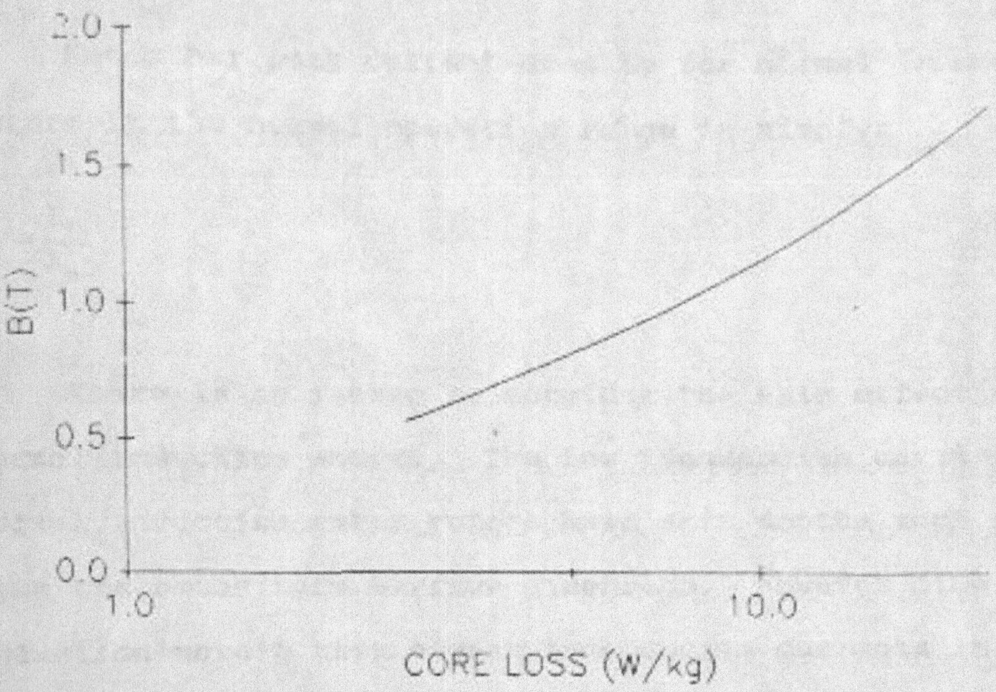

Figure 3 Stator and rotor lamination B versus core loss curve for 26 gauge armature grade .5% silicon steel at 60 Hz. Data from Kuhlmann[42].

4.2.15 Total Input Power, Efficiency, and Power Factor

Total input power is determined by:

$$P_{in} = P_{rt} + 1.5R_sI^2 + P_{Mag} \tag{4-25}$$

Efficiency and power factor are determined by:

$$E_{ff} = \frac{P_{out}}{P_{in}} \tag{4-26}$$

$$PF = \cos(A_{sgvi} - \delta_p) \tag{4-27}$$

4.2.16 Rotor Bar Peak Current Density

Rotor bar peak current density for normal induction motors in the normal operation range is simply:

$$J_r = \frac{I_{rb}}{A_b} \qquad (4\text{-}28)$$

There is no reason to consider the skin effect on normal induction motors. The low frequencies carried by normal induction motor rotors have skin depths much greater than the rotor bars maximum dimension. However high speed induction motors have higher frequencies currents in their rotors. The skin effect should be considered in very high speed induction motors.

4.2.17 Rotor Bar Peak Flux Density

The magnetic flux densities in each rotor bar can be thought of as being caused by two sources, the flux from the rotating field and the self-produced flux caused by the current going through each rotor bar. Each will be considered.

4.2.17.1 Magnetic Flux from the Rotating Field. The upper limit of magnetic flux through the rotor as caused by the rotating field was calculated in equation (4-1). Most of this flux will pass through the lamination material of the rotor, although a small amount will go through the rotor bars.

The rotor bars will experience the same mmf drop across them as the lamination material they are adjacent to. An estimate of the flux density in the rotor bars can be obtained by multiplying the ratio of the reluctance of a flux path through a rotor bar to the reluctance through the adjacent lamination material times the rotor flux density.

When as much lamination material is used as is typical in induction motors the ratio of reluctances is very small. Therefore the flux density in the rotor bars caused by the rotating stator field in the rotor bars is also very small. Calculations have shown it to be small enough to be insignificant in relation to the self-induced rotor bar flux during steady state operation. Self-induced rotor bar flux is calculated in the next section.

4.2.17.2 <u>Magnetic Field from the Rotor Bars</u>. Generally, self-produced flux is not considered in the rating of superconductors. Critical surface curves, such as those shown by Fitzgerald[25], consider externally produced magnetic field. However, because the rotor bars are partially surrounded by laminations, their own currents cause significant flux densities through themselves near the rotor gap.

The maximum magnetic flux density that a rotor bar

forces through itself is proportional to the current going through it and inversely proportional to its slot width. The flux path considered, the radially outer most, has the maximum flux density. It is shown in Figure 4. To simplify calculations the flux path length is approximated as the slot width. This approximation will produce a slightly larger than actual flux density.

$$B_r = \frac{\mu_o I_{rb}}{s_{wr}} \qquad (4\text{-}29)$$

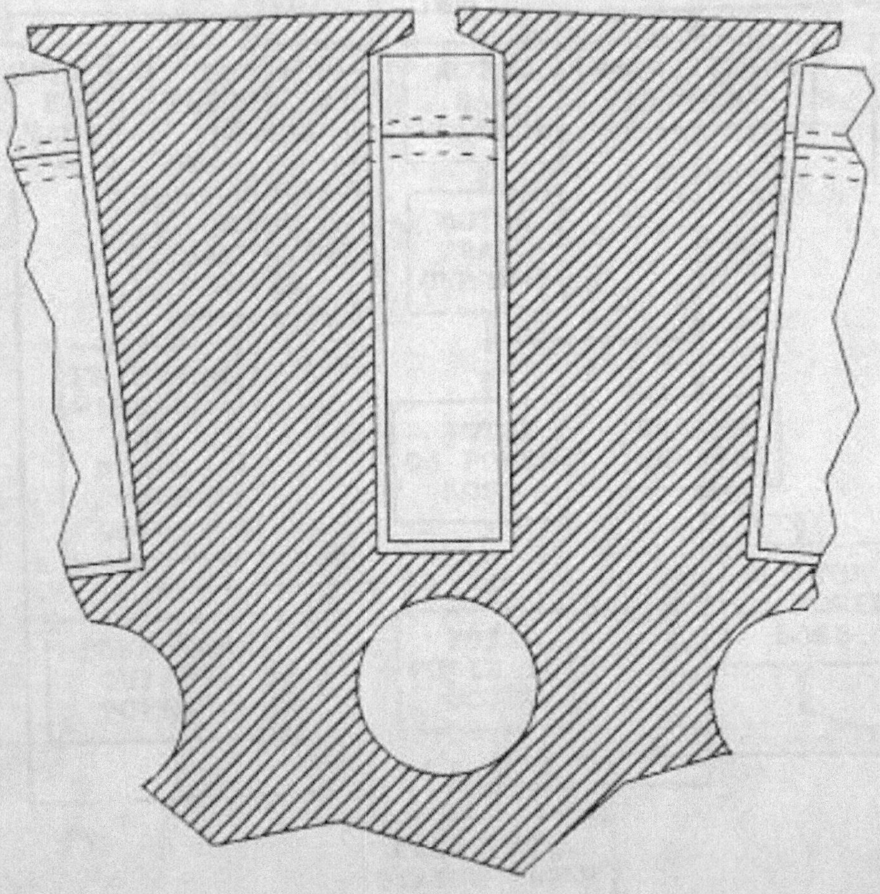

Figure 4 Partial cross-section of the superconducting synchronous/induction motor squirrel cage rotor. The flux paths of the maximum flux density experienced by the rotor bars are shown by dotted lines.

4.2.18 Equation Solution Flowchart

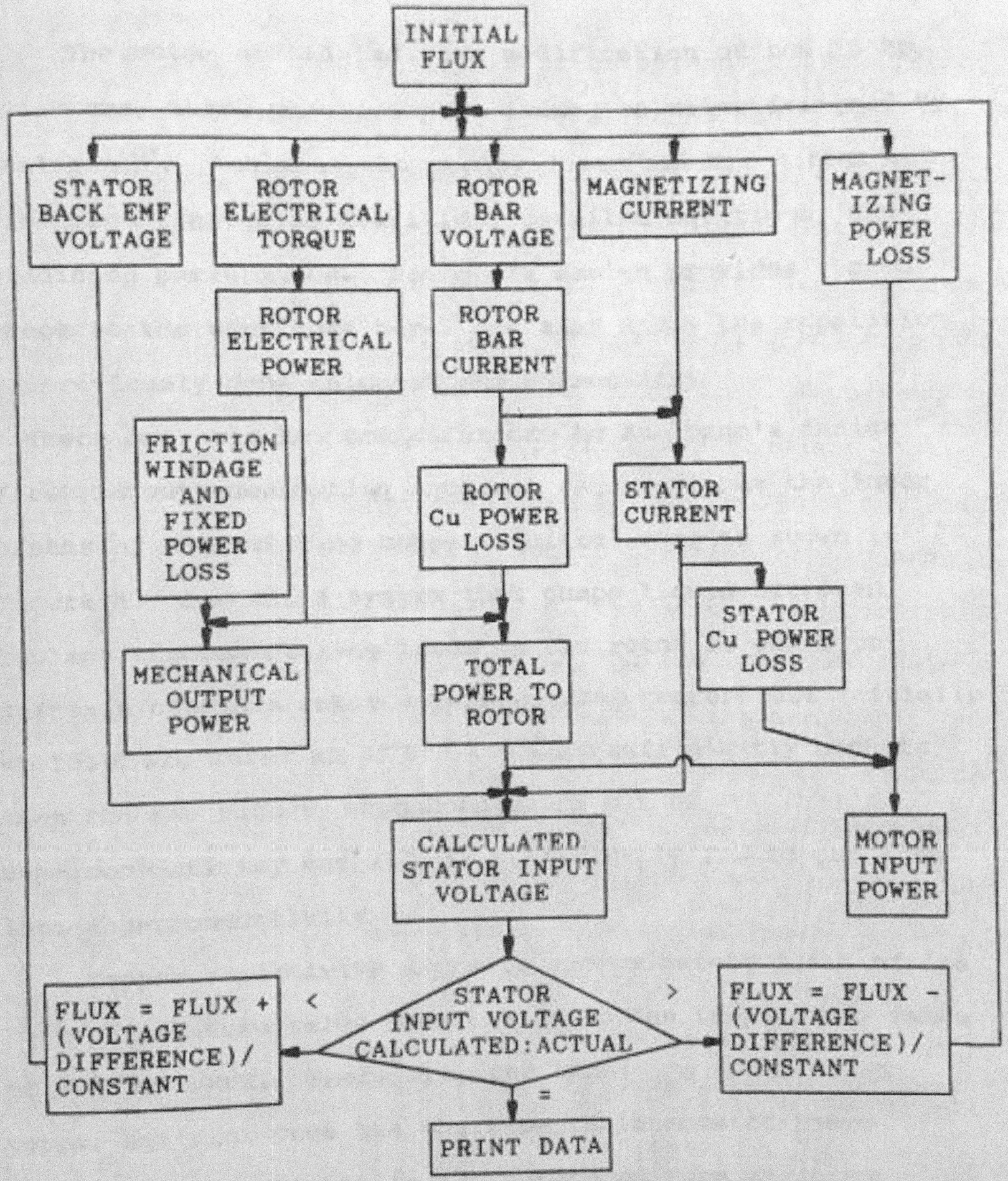

Figure 5 Flowchart showing the order of equations solved with the fixed sinusoidal rotating field method on the superconducting synchronous/induction motor in the steady state, fixed slip, normal conduction (induction motor) mode.

4.3 Applying the Calculation Methods to an Actual Motor

The motor considered is a modification of the 15 hp, 220 V rms, three phase, 6 pole induction motor designed by Kuhlmann[41]. Kuhlmann thoroughly described his design with dimensions, material qualities, detailed equations, and predicted performance. Using his design provides a good check to the work done here. It also makes the repetition of previously done calculations unnecessary.

There are only two modifications to Kuhlmann's design. First, a superconducting squirrel cage replaces the inner 5/6ths of the existing copper squirrel cage as shown in Figure 1. Second, a system that pumps liquid nitrogen coolant through cooling tubes in the rotor is added to maintain constant rotor superconductor temperature initially at 100°K and later at 77°K. 100°K is sufficiently high to keep the new high T_c superconductors out of superconductivity and 77°K is sufficiently low to put them into superconductivity.

Copper resistivity drops to approximately 1/6th of its room temperature value when cooled to the temperature range of 77°K to 100°K. Therefore, the remaining 1/6th sized copper squirrel cage has the same resistance at these temperatures as it had in its unreduced form at room temperature. The superconductor material has a resistance much greater than that of copper before it goes into superconductivity. This results in the total squirrel cage resistivity at liquid nitrogen temperatures before superconductivity being the same as it was in Kuhlmann's room temperature motor.

Induction motor equivalent circuit impedance values given by Kuhlmann were used without modification on the motor during normal conducting starting.

A cross-sectional view of the motor is shown in Figure 1.

4.3.1 Impedances

Kuhlmann reduced his design to a per phase equivalent circuit with all values referred to the stator. The values he calculated for the stator can be used as they are. The rotor values have to be referred back to the rotor and reduced to a perfect pitch coil basis. Leakage inductances are practically constant in the operating range, they will be treated as constants. Magnetizing impedances are not necessary to the method used here. Magnetizing losses are determined from average calculated flux values applied to loss curves. Inductive magnetizing impedance is accounted for with the calculation of magnetizing current.

4.3.1.1 Rotor Resistance. An equation for calculating the resistance of a single rotor bar from the equivalent circuit rotor resistance is given by Fitzgerald and Kingsley[43].

The rotor resistance per perfect pitch two pole coil is:

$$R_{rb} = R_{re} \frac{N_b p}{m(Nk_d k_p)^2} \tag{4-30}$$

$$= .12 \frac{65}{3(162)^2(.985)^2(.96)^2} = 1.108 \times 10^{-4}\ \Omega$$

4.3.1.2 Rotor Inductance. An equation for calculating the inductance of a single rotor bar from the equivalent circuit rotor inductance is given by Fitzgerald and Kingsley[43]. The rotor inductance per perfect pitch two pole coil is:

$$L_{rb} = \frac{X_{re}}{w_s} \frac{N_b p}{m(Nk_d k_p)^2} \tag{4-31}$$

$$= \frac{.42}{377} \cdot \frac{65}{3(162)^2(.985)^2(.96)^2} = 1.029 \times 10^{-6}\ H$$

4.3.2 Per Unit Base Values

Per unit base values were taken from Kuhlmann's 15 hp motor[41].

Variable	Per Unit Base
Power	15 (hp)
Voltage	179.6 (V peak LN)
Rotating speed	1,200 (rpm)
Current (calculated from above)	41.6 (A peak)
Torque (calculated from above)	89.1 (N m)
Moment of inertia (motor moment of inertia)	0.3 (Kg m^2)
Peak superconducting rotor bar current density (calculated at rated slip, copper rotor bar current/superconductor cross-sectional area)	8.77×10^6 (A/m^2)
Peak superconducting rotor bar flux density (calculated at rated slip using copper rotor bar current and outer superconductor magnetic path)	0.125 (T)
Flux per pole (value observed at rated slip)	0.005707 (Wb)
Angle between back emf and source voltage (value observed at rated slip)	0.095 (rad)

4.3.3 FORTRAN Program

The FORTRAN program solves the equations to produce a output torque, peak stator current, peak rotor bar current density, peak rotor bar flux density, efficiency, power factor, and angle between back emf and source voltage. All of these are solved versus slip. The program is run with four rotor resistance values. Rotor resistance values of 1, 1/2, 1/4, and 1/8 times the rotor resistance value calculated for normal operation in section 4.3.1.2.

```
C RFINST.FOR
C THIS ANALYZES KUHLMANN'S MOTOR
C BY CONSIDERING A ROTATING FLUX WAVE
C A TWO POLE MODEL IS USED AND ITS RESULTS CONVERTED
C TO SIX POLE AT THE END OF THE PROGRAM
C 8/28/89
C
C PER UNIT BASE FOR I IS 29.4*SQRT(2) AMPS PEAK
C PER UNIT BASE FOR JR IS 8.77E6 AMPS/M2 PEAK
C PER UNIT BASE FOR BR IS .125 T
C PER UNIT BASE FOR TOUT IS 89.1 NM
        REAL I, IM, JR, IRB, KD, KP, KR, KS, LG, LRB, LS,
     Z          MO, MASS, M, N, NB
                PI = 3.1415927
        MO = PI*4E-7
        ET = 179.62925
        P = 6.
        M = 3.
        NB = 65./P
        N = 162.
        KD = .96
        KP = .985
        WS = 376.99112
        RRB = 1.*.12*(NB*P/(M*(N*KD*KP)**2))
        WRITE(10,*) 'RRB',RRB
        WRITE(10,*) ' '
        LRB = .42/WS*(NB*P/(M*N**2*(KD*KP)**2))
        RS= .15
        LS = .39/WS
        DE = 1.
C DENSITY FOR 4% SILICON-IRON
        DENS= 7650.
        RWS= .0173482
        PFWS= 235.89617
        PFL = 112.
        DEL = 5.59E-4
        LG = .114
        D = .241
        KR = 1.052
        KS = 1.
        CORF = 1.31
        VOL = LG*(PI*((D/2. + RWS)**2 - (D/2.)**2) + 54.*1.087
     Z          *.31*(2.54/100.)**2)
        MASS = DENS*VOL
        AB = (5./6)*.141*.531*(2.54/100.)**2
        SWR = .141*.0254
        WRITE(10,*) ' S(PU)                I(PU)          TOUT(PU)
     Z     EFFICIENCY      POWER FACTOR          PLOSS(WATTS)
     Z   HP                PHI(PU)             JR(PU)
     Z        BR(PU)                     IRB(AMPS PEAK)    OP(PU)'
        WRITE(10,*) ' '
        DO 30 S = .0000001, .1800, .015
C ROTOR CALCULATIONS ARE DONE ON A PER ROTOR COIL BASIS
C THE ROTOR SQUIRREL CAGE IS DIVIDED UP INTO 3 PHASE GROUPS
```

```
C EACH PHASE GROUP CONTAINS 65/6 PERFECT PITCH ONE TURN COILS
C STATOR CALCULATIONS ARE DONE ON A PER PHASE BASIS
C POWERS AND TORQUES ARE TOTAL VALUES
C FLUX IS THE FLUX COMING FROM EACH POLE
          PHI = 2.*ET/KD/KP/WS/N
C ROTOR PHASE VOLTAGE IS CALCULATED FROM THE TOTAL FLUX
C IT IS THE VOLTAGE ON EACH ROTOR COIL
10        ER = PHI*WS*S
C STATOR PHASE VOLTAGE AT THE AIR GAP IS CALCULATED FROM THE
C TOTAL FLUX
C IT IS THE VOLTAGE PRODUCED BY EACH PHASE
          ETG = PHI*KD*KP*WS*N/2.
C ROTOR PHASE CURRENT IS CALCULATED FROM THE ROTOR
C VOLTAGE
          IRB = ER/2./(RRB**2 + (LRB*WS*S)**2)**.5
C TORQUE PRODUCED BY THE ACTION OF THE ROTOR CURRENTS AGAINST
C THE ROTOR CURRENTS.  ALL PHASES HAVE BEEN COMBINED HERE.
C
          TE = .375*NB*P/M*PHI**2*WS*S*RRB/(RRB**2+(LRB*WS*S)**2)
C POWER DEVELOPED BY THE ROTOR TORQUE
          PE = TE*WS*(1 - S)
C TOTAL POWER LEAVING THE STATOR
          PRT = PE + NB*P/2.*RRB*IRB**2
C MAGNETIZING CURRENT
          IM = PHI*1.247E6*P**2*DEL*KR*KS/M/N/KD/KP/D/LG*CORF
C STATOR CURRENT IS A TURNS RATIO TIMES THE ROTOR CURRENT
C PLUS A MAGNETIZING CURRENT
                    I = ((IM + IRB*NB*P*LRB*WS*S/(M*N*KD*KP*(RRB**2 +
     Z                  (LRB*WS*S)**2)**.5))**2 + (IRB*NB*P*RRB/
     Z                  (M*N*KD*KP*(RRB**2 + (LRB*WS*S)**2)**.5))**2)
     Z                  **.5
C ANGLE BETWEEN THE STATOR VOLTAGE AND CURRENT AT GAP
          ASGVI = -ACOS(PRT/I/ETG/1.5)
C REAL PART OF THE CALCULATED STATOR VOLTAGE
          ET1R = ETG + I*(RS**2 + (LS*WS)**2)**.5*
     Z                  COS(ASGVI + ATAN(LS*WS/RS))
C IMAGINARY PART OF THE CALCULATED STATOR VOLTAGE
          ET1I = I*(RS**2 + (LS*WS)**2)**.5*
     Z                  SIN(ASGVI + ATAN(LS*WS/RS))
C MAGNITUDE OF THE CALCULATED STATOR VOLTAGE
          ET1 = (ET1R**2 + ET1I**2)**.5
C ANGLE OF THE CALCULATED STATOR VOLTAGE RELATIVE TO THE VOLTAGE
C AT THE AIR GAP
          DP = ATAN(ET1I/ET1R)
C DIFFERENCE BETWEEN THE MAGNITUDE TO THE CALCULATED STATOR INPUT
C VOLTAGE AND THE ACTUAL INPUT VOLTAGE
          DE = ET - ET1
C REPEATING THE PROGRAM IF THE DIFFERENCE IS TOO GREAT
          IF (ABS(DE).GT..1) THEN
          PHI = PHI + DE/KD/KP/N/WS
          GOTO 10
          ENDIF
C CALCULATING THE ACTUAL OUTPUT POWER OF THE MOTOR
                POUT = PE - (1 - S)**1.35*PFWS - PFL
```

```
          HP = POUT/746
C CALCULATING THE ACTUAL OUTPUT TORQUE OF THE MOTOR
                 TOUT = 3.*POUT/(1 - S)/WS
C CALCULATING THE STATOR PEAK FLUX DENSITY
          BS = PHI/2./RWS/LG
                 PMAG = 10**(BS/1.1)*MASS
C CALCULATING THE TOTAL INPUT POWER
                 PIN = PRT + 1.5*I**2*RS + PMAG
C CALCULATING THE POWER LOSSES
          PLOSS = PIN - POUT
C CALCULATING THE OVERALL EFFICIENCY
          EFF = POUT/PIN
C CALCULATING POWER FACTOR
          PF = COS(ASQVI - DP)
C CALCULATING ROTOR BAR CURRENT DENSITY
          JR = IRB/AB
C CALCULATING ROTOR FLUX
          BR = IRB*MO/SWR
C
C CALCULATING PER UNIT VALUES
          I = I/29.4/1.4142136
          PUJR = JR/8.7766
          PUBR = BR/.125
          PUTOUT = TOUT/89.1
          PUPHI = PHI/5.707E-3
          PUDP = DP/.095
          WRITE(10,*) S,I,PUTOUT,EFF,PF,PLOSS,HP,PUPHI,
     Z                 PUJR,PUBR,IRB,PUDP
          WRITE(10,*) ' '
30        CONTINUE
          END
```

4.3.4 Graphed FORTRAN Program Results

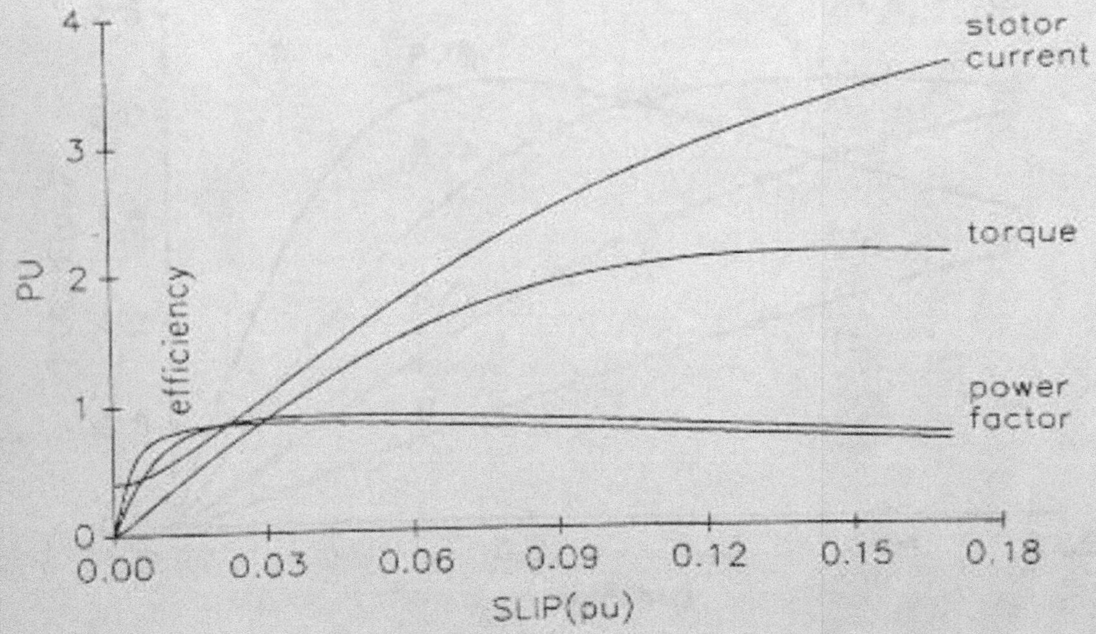

Figure 6 Characteristics of the superconducting synchronous/induction motor in the steady state normal conducting induction motor mode. Given are per unit stator current, torque, power factor, and efficiency versus slip. Per unit bases are given in 4.3.2.

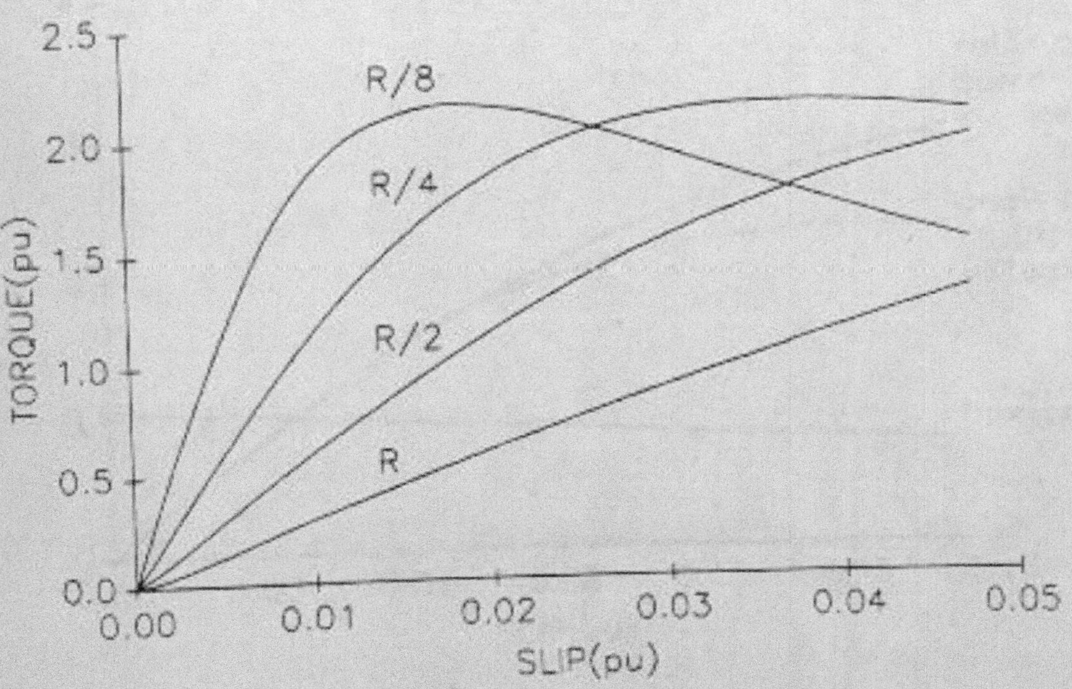

Figure 7 Torque versus slip for the superconducting synchronous/induction motor in the steady state normal conducting induction motor mode for a range of effective rotor bar resistances. Effective rotor bar resistance is dropped as parts, but not all, of the squirrel cage become superconductive. R is the perfect pitch rotor bar resistance used by Kuhlmann's induction motor, 1.108 x 10^{-4} Ω. The per unit base for torque is 89.1 Nm.

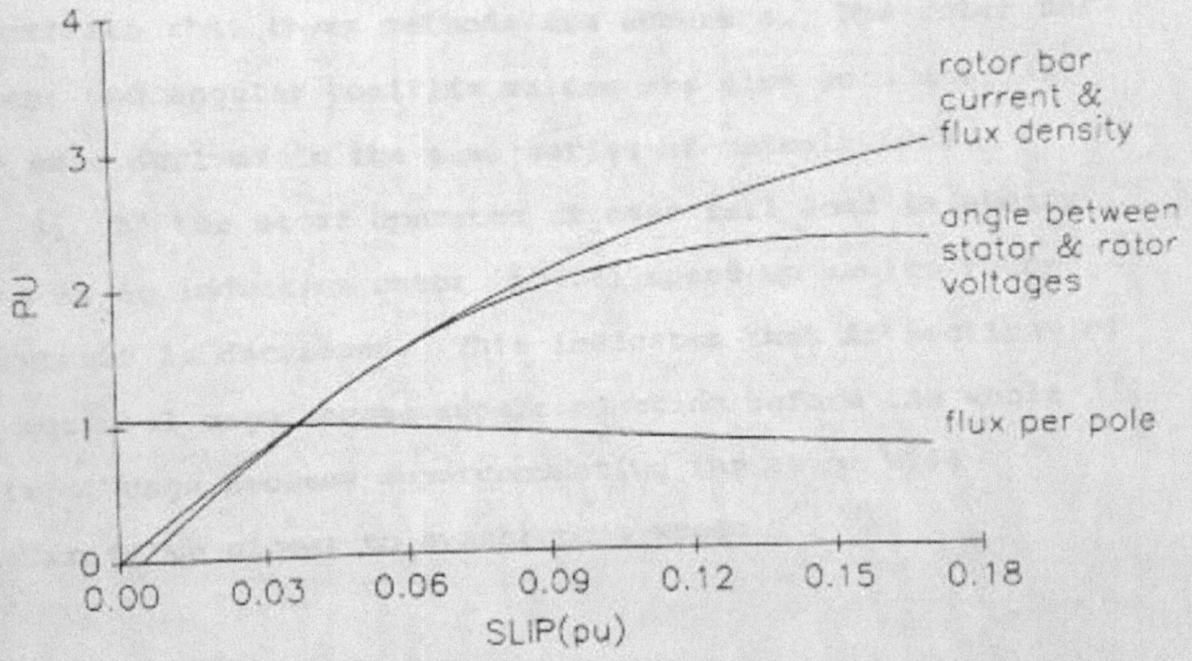

Figure 8 For the superconducting synchronous/induction motor in the steady state normal conducting induction motor mode, maximum current density, and magnetic flux density that could be experienced in the superconducting portions of the rotor bars versus slip. The maximums are those that would occur if the superconducting portion of the rotor bars assumed the maximum value carried in the outer copper rotor bar portion. Rotor bar resistance is the effective perfect pitch resistance used by Kuhlmann's induction motor, $1.108 \times 10^{-4}\ \Omega$. Per unit bases are given in 4.3.2.

4.3.5 Observations and Conclusions

1) The computation methods used on the example 15 hp motor produce results consistent with Kuhlmann's[41]. The agreement with Kuhlmann's motor characteristics and the care that was put into setting up the computation procedure demonstrate that these methods are accurate. The rotor bar current and angular position values are also accurate, as they were derived in the same series of calculations.

2) If the motor operates at near full load in steady state as an induction motor it will speed up as its rotor resistance is decreased. This indicates that if sections of the squirrel cage become superconducting before the whole squirrel cage becomes superconducting the rotor will accelerate to closer to synchronous speed.

4.3.6 Program Output Data, Superconducting Synchronous/Induction Motor in the Normal Conduction Mode

RRB 1.1079710E-04

S(PU)	I(PU)	TOUT(PU)	EFFICIENCY	POWER FACTOR
PLOSS(WATTS)	HP	PHI(PU)	JR(PU)	BR(PU)
IRB(AMPS PEAK)	DP(PU)			
1.0000000E-07	0.3953450	-3.1050276E-02	-0.5837219	1.3728382E-02
943.2480	-0.4660303	1.051547	2.8920767E-06	2.8658685E-06
1.0208612E-03	-0.1444217			
1.5000099E-02	0.6559008	0.4498890	0.8196291	0.7554533
1091.888	6.651051	1.029848	0.4242783	0.4204335
149.7788	0.3321753			
3.0000100E-02	1.094202	0.8740947	0.8580914	0.8645175
1569.976	12.72562	1.003131	0.8231527	0.8156931
290.5894	0.7590076			
4.5000099E-02	1.528697	1.229002	0.8510964	0.8786348
2299.164	17.61589	0.9729478	1.189497	1.178718
419.9163	1.127648			
6.0000099E-02	1.929061	1.511506	0.8327236	0.8659486
3195.648	21.32488	0.9410221	1.519714	1.505942
536.4898	1.434693			
7.5000100E-02	2.288198	1.725599	0.8103172	0.8465464
4183.523	23.95688	0.9088022	1.813187	1.796756
640.0914	1.680931			
9.0000100E-02	2.605749	1.879462	0.7863776	0.8231475
5202.092	25.66986	0.8773829	2.071429	2.052657
731.2559	1.870165			
0.1050001	2.884209	1.982970	0.7619706	0.7980020
6207.531	26.63716	0.8475075	2.297237	2.276419
810.9706	2.008034			
0.1200001	3.128281	2.047274	0.7376181	0.7723902
7175.438	27.04004	0.8198745	2.494782	2.472174
880.7082	2.101047			
0.1350001	3.339985	2.078438	0.7135754	0.7470880
8079.994	26.98372	0.7941296	2.665925	2.641766
941.1249	2.155868			
0.1500001	3.524483	2.085578	0.6899948	0.7225663
8917.766	26.60688	0.7706503	2.814936	2.789427
993.7289	2.178856			
0.1650001	3.685520	2.074857	0.6669538	0.6990981
9686.588	26.00299	0.7493829	2.944907	2.918220
1039.611	2.175805			

5.0 MOTOR WITH A SUPERCONDUCTING SQUIRREL CAGE ROTOR MODELLED WITH A FIXED SINUSOIDAL ROTATING FIELD TO ESTIMATE SUPERCONDUCTOR REQUIREMENTS

The same superconducting synchronous/induction motor shown in Figure 1 and analyzed in chapter 4.0 is analyzed again. Here the superconductor material is considered in the superconducting state. The superconducting rotor bars form complete superconducting circuits and trap flux. This chapter estimates the superconductor material characteristics required for the superconductor to stay superconducting once all the material has made the transition to superconductivity.

5.1 Approximations and Parameters

The same approximations and parameters applied in 4.0 apply here, except that here a superconductive squirrel cage is considered. There are several significant assumptions that must be made about the superconductive squirrel cage. First, the squirrel cage is totally superconductive during all times considered. This means it contains a fixed magnitude and distribution of magnetic flux. This avoids the analysis problems that would exist if only some of the rotor bars were superconductive, as then trapped flux could exist is just parts of the rotor. It would be very difficult to realistically predict where the trapped flux might occur and to predict its subsequent effects on the motor performance.

Details of the rotor cooling system are not considered here.

Stator resistance and leakage impedance are assumed constant in this analysis. The stator resistance is the same as it was in chapter 4.0. Lowering stator winding resistance will not be considered here. The leakage impedance is not the same as it was in chapter 4.0. The total flux leakage should be about the same here as it was in chapter 4.0. Calculating that leakage from basics is not easy. It is as difficult here as it was for the induction motor of chapter 4.0. However, by assuming the same total leakage flux occurs here as in 4.0 the effective stator input leakage inductance can be taken as simply the sum of the induction motor's stator leakage inductance and the reflected rotor's leakage inductance.

Rotor resistance is zero since the rotor bars are superconductive.

5.2 Steady State Operation

Steady state operation for the superconducting synchronous/induction motor means the rotor is revolving at synchronous speed, much like a synchronous motor operating at steady state.

Motor performance is determined at different power angles for varying amounts of motor load.

Rotor flux densities and distributions of flux are taken from chapter 4.0, the flux immediately prior to superconductivity.

5.2.1 Equations

Rotor flux is constant in magnitude and distribution. During steady state it also rotates at a constant speed. Except for slight flux leakages the rotor flux also forces a constant flux through the stator. An effective back emf is induced on the stator windings by the rotating rotor flux.

The equations have been written for a 2 pole motor. However, they can be applied to motors with more poles by applying appropriate multiplication factors.

All equations use peak values unless otherwise stated.

5.2.1.1 Stator Current. The stator current is solved for with the phasor equation:

$$I = \frac{E_T - .5Nw_s k_d k_p \phi_M \angle \delta_p}{R_s + w_s(L_{re} + L_s)j} \qquad (5\text{-}1)$$

5.2.1.2 Power to the Motor. Total power into the motor is:

$$P_{in} = 1.5|E_T I^*| \qquad (5\text{-}2)$$

5.2.1.3 Magnetic, Friction, Fixed, and Resistive Losses. There are hysteresis and eddy current losses in the stator as there were during induction motor operation. The rotor with its fixed magnetic field has no significant magnetic losses. The losses are calculated as they were in chapter 4.2.14.

Friction losses are constant during synchronous operation. They are P_{fws}.

Fixed losses are P_{fl}.

Only the stator has resistive losses. They are calculated as:

$$\text{Resistive Losses} = 1.5R_sI^2 \qquad (5\text{-}3)$$

5.2.1.4 Output Power, Torque, Efficiency, and Power Factor. The equation for output power is:

$$P_{out} = P_{in} - P_{Mag} - 1.5R_sI^2 - P_{fws} - P_{fl} \qquad (5\text{-}4)$$

The equation for output torque is:

$$T_{out} = \frac{P_{out}}{W_s} \qquad (5\text{-}5)$$

The equation for efficiency is:

$$E_{ff} = \frac{P_{out}}{P_{in}} \qquad (5\text{-}6)$$

Power factor is the cosine of the angle between I and E_r.

5.2.1.5 <u>Steady State Output Performance Equation Solution Flowchart</u>.

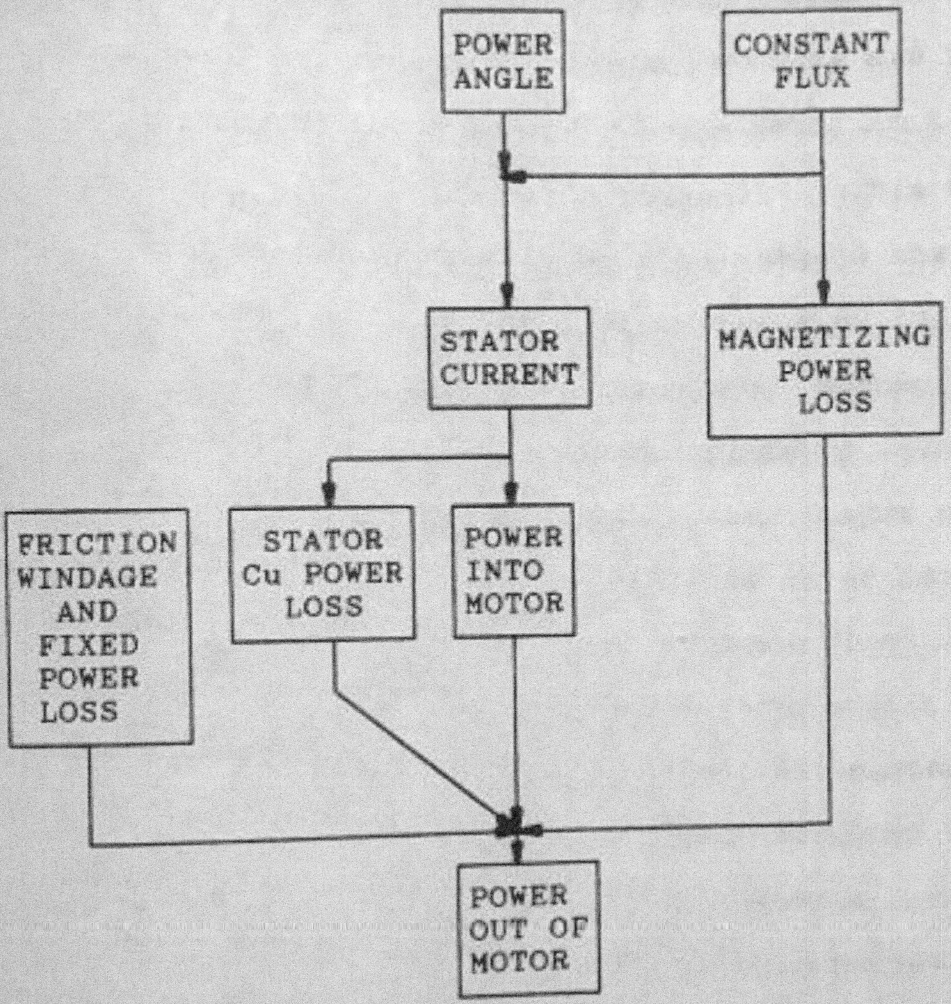

Figure 9 Flowchart showing the order of equations solved with the fixed sinusoidal rotating field method on the superconducting synchronous/induction motor in the steady state superconduction (synchronous motor) mode.

5.2.1.6 Rotor Bar Current. Immediately prior to the transition to superconductivity rotor bar current magnitudes range from zero to the peak value calculated in equation (4-3). After the transition these currents are superimposed by reflected stator currents. As the motor is loaded its power angle and stator currents increase. This causes the stator to attempt to force more flux through the rotor. The rotor with its trapped flux opposes the flux increase by creating currents in its rotor bars that oppose the increase. The rotor bar currents initially maintaining the "trapped flux" and those countering the stator currents combine to result in an increased peak rotor bar current.

To determine the rotor bar currents first consider the portion induced by stator currents. The stator conductors will act as a current sheet that will be countered by an opposing rotor current sheet. Figure 10 shows the A phase of a two pole three phase motor producing a current sheet and the rotor bars producing an opposing current sheet. The location of the current sheet is fixed in the stator frame of reference as shown in Figure 10. All three phases produce currents simultaneously. The approximate currents induced into the rotor bars versus a counterclockwise angle around the stator are:

$$I_{rs}(\theta_s) = \frac{INm}{N_b p}\left[\sin(w_s t)K(\theta_s) + \sin\left(w_s t + \frac{2\pi}{3}\right)K\left(\theta_s + \frac{2\pi}{3}\right) + \sin\left(w_s t - \frac{2\pi}{3}\right)K\left(\theta_s + \frac{\pi}{3}\right)\right] \quad (5\text{-}7)$$

Where $K(\theta_s)$ is described in the equation:

$$K(\theta_s) = \begin{cases} 1 & \pi/6 \geq \theta_s > -\pi/6 \\ 0 & 5\pi/6 \geq \theta_s > \pi/6 \\ -1 & 7\pi/6 \geq \theta_s > 5\pi/6 \\ 0 & 11\pi/6 \geq \theta_s > 7\pi/6 \end{cases} \quad (5\text{-}8)$$

It should be mentioned that equations (5-7) and (5-8) predict discontinuous current changes that would not occur in an actual motor. The actual current changes would not be discontinuous, currents would rise and drop in finite times. However, it is felt that the equations adequately describe the major portions of the current waveform.

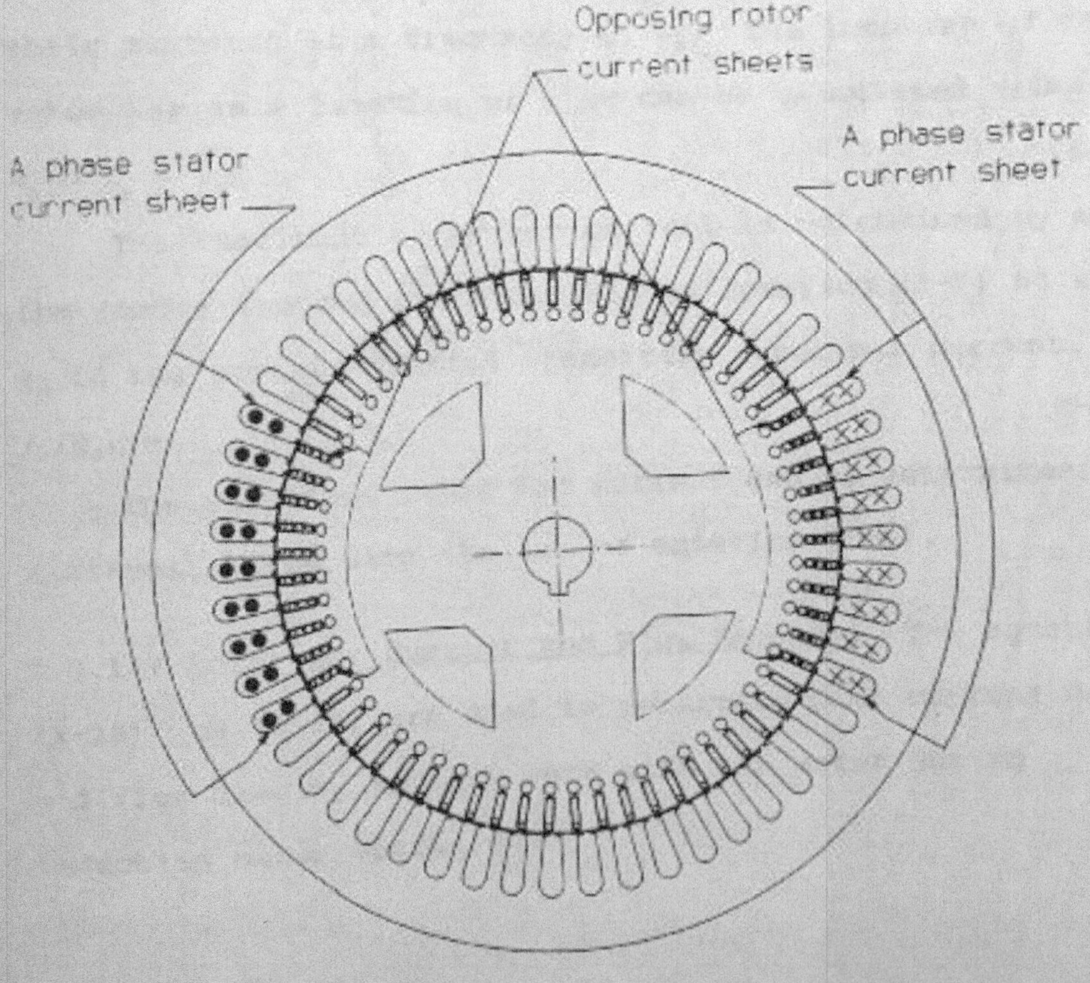

Figure 10 Stator and rotor current sheets during superconduction.

The most significant rotor bar for this analysis is the one carrying the peak current at the time of transition. It will be exposed to the greatest currents and magnetic fields. The position of this rotor bar is δ_p radians counterclockwise from the A phase coil center coil at time t = 0.0. As time progresses the rotor rotates clockwise at the constant rate of w_s while the stator windings change their currents at a frequency of w_s. The location of the rotor bar as a function of time can be calculated with:

$$\theta_s = \delta_p - w_s t \tag{5-9}$$

The resultant rotor bar current is determined by adding the stator induced rotor current of equation (5-7) at angle θ_s to the maximum initial transition rotor bar current.

$$I_r(\theta_s) = I_{rbi} + I_{rs}(\theta_s) \tag{5-10}$$

The resultant rotor bar current can be determined for different times with the use of equation (5-9).

5.2.1.7 Rotor Bar Current and Flux Density. The equations (4-28) and (4-29) are used to determine peak current density and flux density as they were with the motor during induction motor operation.

5.2.1.8 Steady State Flux and Current Density Equation Solution Flowchart.

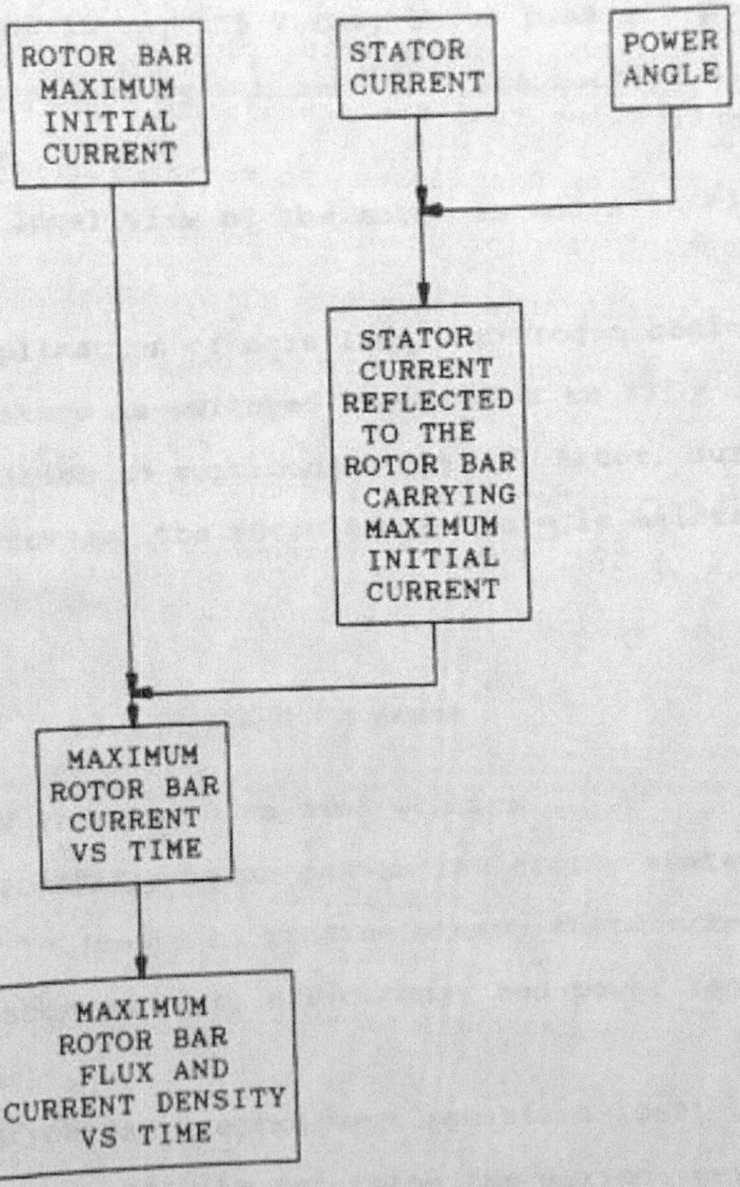

Figure 11 Flowchart showing the order of equations solved to determine the maximum rotor bar flux and current density waveform. This applies to the fixed sinusoidal rotating field method on the superconducting synchronous/induction motor in the steady state superconduction (synchronous motor) mode.

5.3 Applying the Calculation Methods to an Actual Motor

The motor considered is as considered in section 4.3, a modification of the 15 hp, 220 V rms, three phase, 6 pole induction motor designed by Kuhlmann[41]. See section 4.3 for a description.

A cross-sectional view of the motor is shown in Figure 1.

With the application of more liquid nitrogen coolant the rotor temperature is switched from 100° K to 77° K during the transition to superconductivity. Later, during steady state operation, the rotor temperature is maintained continuously at 77° K.

5.4 FORTRAN Programs

Two FORTRAN programs have been written.

The first FORTRAN program solves the steady state equations (5-1) to (5-6) to produce steady state output torque, peak stator current, efficiency, and power factor versus power angle.

The second FORTRAN program uses equations (5-7) to (5-10) and (4-28) to (4-29) to determine the maximum current density and magnetic flux density waveform experienced by the rotor bars.

5.4.1 FORTRAN Program Describing the Superconducting Synchronous/Induction Motor in the Steady State Superconducting Mode

```
C SUPERCONDUCTING INDUCTION MOTOR DURING STEADY STATE OPERATION
C KUHLMANN'S MOTOR IS ANALYZED WITH THE ASSUMPTIONS OF SECTION 5.0
C
C 4/25/89
C
            REAL I, II,  KD, KP, K1,
     Z           K2, K3, LG, LSY, M, MASS, MO, N, NB
C
            PI = 3.1415927
            MO = PI*4.E-7
            KD = .96
            KP = .985
            D = .241
            RWS = .0173482
            WS = 376.8
            LG = .114
            RSY = .15
            LSY = (.39 + .42)/WS
            P = 6.
            N = 162.
            NB = 66./P
            M = 3.
            ET = 179.62925
            PHI = .0057072212
            PFWS = 232.
            PFL = 112.
            DENS = 7650.
            VOL = LG*(PI*((D/2. + RWS)**2 - (D/2.)**2) + 54.*1.067
     Z       *.31*(2.54/100.)**2)
            AS = 5./6.*.141*.531*.0254**2
            SWR = .141*.0254
            MASS = DENS*VOL
            WRITE(6,*) ' DP(RAD)           I(PU)            TOUT(PU)
     Z       EFFICIENCY      POWER FACTOR          PLOSS(WATTS)      HP'
            WRITE(6,*) ' '
            DO 10 DP = 0.0, -.40, -.05
C ALL STATOR CALCULATIONS ARE PER PHASE
C PEAK VOLTAGES AND CURRENTS ARE USED
C
C DETERMINING THE STATOR CURRENT
            A = ET - .5*N*WS*PHI*KD*KP*COS(DP)
            B = -.5*N*WS*PHI*KD*KP*SIN(DP)
            RI = (A*RSY + B*(LSY*WS))/(RSY**2 + (LSY*WS)**2)
            II = (B*RSY - A*(LSY*WS))/(RSY**2 + (LSY*WS)**2)
            I = (RI**2 + II**2)**.5
C DETERMINING THE POWER FACTOR
            PF = RI/I
C MAGNETIC LOSSES IN THE STATOR
            BS = PHI/2./RWS/LG
            PMAG = 10.**(BS/1.1)*MASS
C TOTAL INPUT POWER
            PIN = 1.5*ET*I*PF
C TOTAL LOSSES
            PLOSS = PMAG + PFWS + 1.5*RSY*I**2 + PFL
C TOTAL OUTPUT POWER
            POUT = PIN - PLOSS
C OUTPUT TORQUE
            TOUT = POUT*P/WS/2.
C EFFICIENCY
            EFF = POUT/PIN
            HP = POUT/746.
C CONVERTING TO OUTPUT TO PER UNIT
            PUI = I/29.4/1.414214
            PUTOUT = TOUT/89.1
            WRITE(6,*) DP,PUI,PUTOUT,EFF,PF,PLOSS,HP
            WRITE(6,*) ' '
 10         CONTINUE
            END
```

5.4.2 FORTRAN Program Evaluating the Maximum Rotor Bar Current Waveform for the Superconducting Synchronous/Induction Motor in the Steady State Superconducting Mode

```
$ type motor10.for
C DETERMINING THE WAVEFORM FOR THE CURRENTS IN A SINGLE
C ROTOR BAR IN A SUPERCONDUCTING INDUCTION MOTOR DURING STEADY
C STATE SUPERCONDUCTING OPERATION
C ONE ROTOR REVOLUTION IS CONSIDERED
C KUHLMANN'S MOTOR IS ANALYZED WITH THE ASSUMPTIONS OF SECTION 5.0
C
C 5/15/89
C
                REAL I, IRB, IRBI, K1, K2, K3, M, N, NB
C
                AB = 4.025315E-5
                WS = 376.8
                N = 182.
                P = 6.
                NB = 66./P
                M = 3.
                IRB1 = 356.
                DP = .2
                I = 45.4
C
                WRITE(6,*) 'DP =',DP
                WRITE(6,*) 'I =',I
                WRITE(6,*) ' '
                WRITE(6,*) '  TI(SEC)        JR(PU)'
                DO 35 TI = 0.0, .017, .0002
                     THS = 6.2831852 - WS*TI + DP
                     IF (THS.GT.6.2831853) THEN
                       A = THS/6.2831853
                       B = INT(A)
                       C = A - B
                       THS = C*6.2831853
                      ENDIF
                     IF (((THS.GE.5.759587).AND.(THS.LT.6.2831853)).OR.
                         ((THS.GE.0.0).AND.(THS.LT..523598))) THEN
     Z
                         K1 = 1.0
                         K2 = 0.0
                         K3 = 0.0
                     ENDIF
                     IF ((THS.GE..523598).AND.(THS.LT.1.570796)) THEN
                         K1 = 0.0
                         K2 = -1.0
                         K3 = 0.0
                      ENDIF
                     IF  ((THS.GE.1.570796).AND.(THS.LT.2.6179939)) THEN
                         K1 = 0.0
                         K2 = 0.0
                         K3 = -1.0
                     ENDIF
                     IF ((THS.GE.2.6179939).AND.(THS.LT.3.665191)) THEN
                         K1 = -1.0
                         K2 = 0.0
                         K3 = 0.0
                      ENDIF
                     IF  ((THS.GE.3.665191).AND.(THS.LT.4.712389)) THEN
                         K1 = 0.0
                         K2 = 1.0
                         K3 = 0.0
                     ENDIF
                     IF ((THS.GE.4.712389).AND.(THS.LT.5.7595865)) THEN
                         K1 = 0.0
                         K2 = 0.0
                         K3 = 1.0
                     ENDIF
                   IRB = IRBI + I*N*M/NB/P*(SIN(WS*TI)*K1
                         + SIN(WS*TI + 2.0943951)*K2
     Z                   + SIN(WS*TI - 2.0943951)*K3)
     Z          PUJR = IRB/AB/2.77E6
                WRITE(6,*) TI,PUJR
                 CONTINUE
35              END
```

5.5 Graphed Program Results

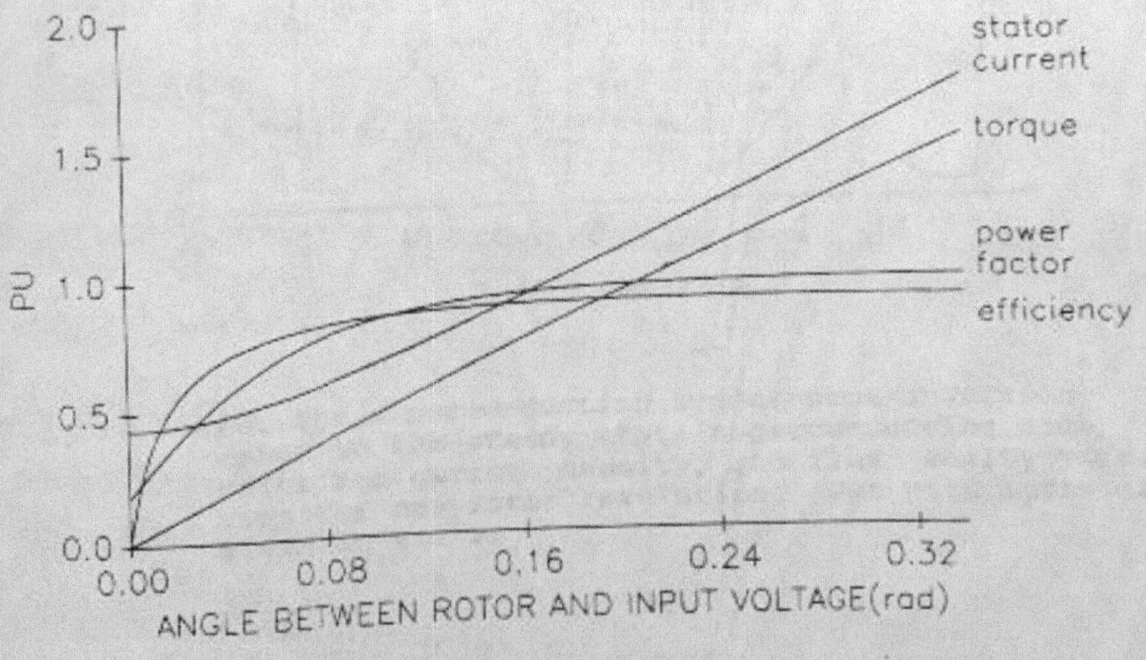

Figure 12 Characteristics of the superconducting synchronous/induction motor in the steady state superconducting mode. Given are per unit stator current, torque, power factor, and efficiency versus angle between the rotor and input voltage. Per unit bases are given in 4.3.2.

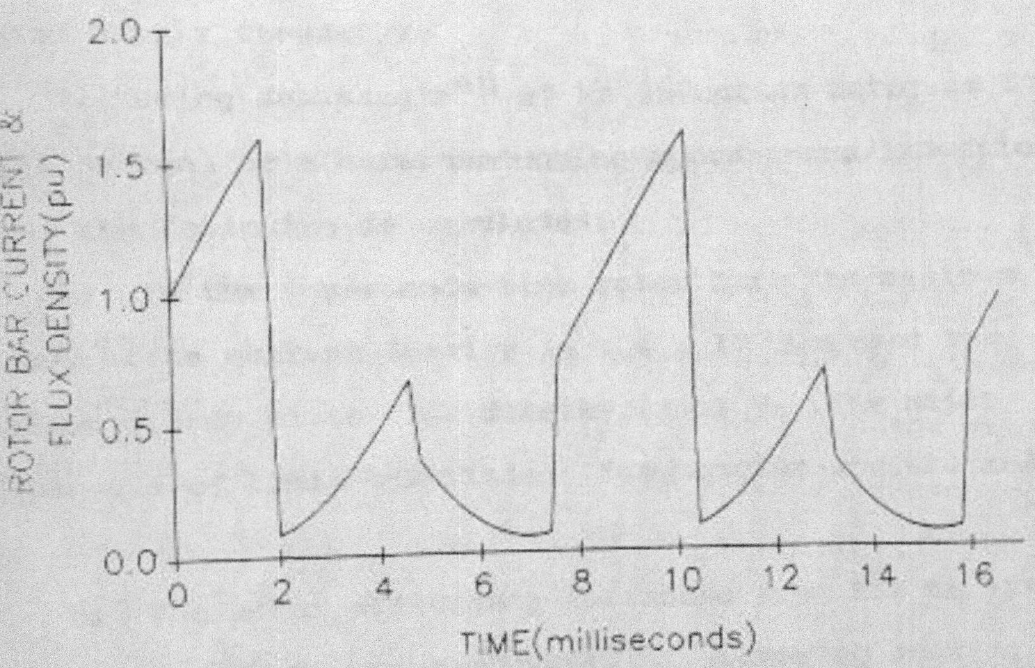

Figure 13 For the superconducting synchronous/induction motor in the steady state superconducting mode, rotor bar current density, and flux density versus time for one rotor revolution. Per unit bases are given in 4.3.2.

5.6 Observations and Conclusions

1) In a superconducting synchronous/induction motor a superconductor material must be able to withstand an AC current density and flux density waveform. The frequencies of the major peaks of the waveform are 2 and 4 times the stator supply frequency.

2) Using Kuhlmann's[41] 15 hp induction motor as the basic design for a superconducting synchronous/induction motor the following is concluded:

a) In the superconducting rotor bars the maximum steady state current density is 1.4×10^7 A/m^2 and the maximum steady state flux density is .2 T. The major components of these quantities' frequencies are 120 and 240 Hz.

b) The motor efficiency increased from 87% to 92% upon being converted to superconductivity, ignoring cooling losses.

c) The motor power factor increased from 87% to 95% upon being converted to superconductivity.

5.7 Program Output Data

5.7.1 Superconducting Synchronous/Induction Motor in the Steady State Superconducting Mode

RUN MOTOR8.EXE

DP(RAD) PLOSS(WATTS)	I(PU) HP	TOUT(PU)	EFFICIENCY	POWER FACTOR
0.0000000E+00 875.4863	0.4355003 1.7296758E-02	1.1530183E-03	1.4524462E-02	0.1820893
-2.5000000E-02 881.6172	0.4532359 1.788290	0.1192091	0.6021011	0.4363674
-5.0000001E-02 900.0063	0.5025930 3.549958	0.2366436	0.7463539	0.6300526
-7.5000003E-02 930.6423	0.5757287 5.301201	0.3533831	0.8095030	0.7574361
-0.1000000 973.5056	0.6645771 7.040923	0.4693546	0.8436396	0.8362485
-0.1250000 1028.570	0.7636945 8.768034	0.5844855	0.8641170	0.8847450
-0.1500000 1095.800	0.8695258 10.48146	0.6987042	0.8770831	0.9151803
-0.1750000 1175.155	0.9798427 12.18013	0.8119390	0.8854796	0.9348137
-0.2000000 1266.584	1.093230 13.86298	0.9241194	0.8908904	0.9478283
-0.2250000 1370.032	1.208763 15.52896	1.035175	0.8942439	0.9566498
-0.2500000 1485.432	1.325819 17.17702	1.145036	0.8961200	0.9627323
-0.2750000 1612.712	1.443962 18.80614	1.253635	0.8968992	0.9669592
-0.3000000 1751.795	1.562882 20.41530	1.360903	0.8968416	0.9698886
-0.3250000 1902.591	1.682346 22.00349	1.466774	0.8961309	0.9718806
-0.3500001 2065.007	1.802180 23.56972	1.571180	0.8949000	0.9731728

5.7.2 Maximum Rotor Bar Current for the Superconducting Synchronous/Induction Motor in the Steady State Superconducting Mode

```
LINK /EXE=MOTOR10.EXE MOTOR10.OBJ
RUN MOTOR10.EXE

DF =  0.2000000
I =   45.40000

  T(SEC)           JR(PU)
 0.0000000E+00    1.008441
 3.9999999E-04    1.150633
 7.9999998E-04    1.289600
 1.1999999E-03    1.422193
 1.6000000E-03    1.545405
 1.9999999E-03   8.6381860E-02
 2.3999999E-03   0.1292617
 2.7999999E-03   0.1920559
 3.1999999E-03   0.2733704
 3.5999999E-03   0.3713515
 3.9999997E-03   0.4837778
 4.3999995E-03   0.6080999
 4.7999993E-03   0.3550450
 5.1999991E-03   0.2595282
 5.5999989E-03   0.1809918
 5.9999987E-03   0.1212169
 6.3999984E-03   8.1558116E-02
 6.7999982E-03   6.2915273E-02
 7.1999980E-03   6.5710798E-02
 7.5999978E-03   0.7486328
 7.9999976E-03   0.8883137
 8.3999978E-03    1.030719
 8.7999981E-03    1.172619
 9.1999983E-03    1.310796
 9.5999986E-03    1.442118
 9.9999988E-03    1.563606
 1.0399999E-02   9.1715917E-02
 1.0799999E-02   0.1377732
 1.1200000E-02   0.2035717
 1.1600000E-02   0.2876195
 1.2000000E-02   0.3880108
 1.2400000E-02   0.5024696
 1.2800001E-02   0.6284005
 1.3200001E-02   0.3390996
 1.3600001E-02   0.2461000
 1.4000001E-02   0.1703855
 1.4400002E-02   0.1136725
 1.4800002E-02   7.7247255E-02
 1.5200002E-02   6.1935477E-02
 1.5600002E-02   6.8084382E-02
 1.6000003E-02   0.7701296
 1.6400002E-02   0.9104468
 1.6800001E-02    1.052986
```

6.0 EQUIVALENT CIRCUIT METHODS FOR DETERMINING SUPERCONDUCTING SYNCHRONOUS/INDUCTION MOTOR PERFORMANCE

Another way of determining the performance of superconducting synchronous/induction motors is with equivalent circuits. Initially, before superconduction takes place, superconducting synchronous/induction motors are easily modelled with the standard induction motor steady state phasor equivalent circuit shown in Figure 14. Later, after their rotor bars have become superconducting and they are in steady state synchronous operation, they are modelled with the fixed field synchronous motor equivalent circuit shown in Figure 15.

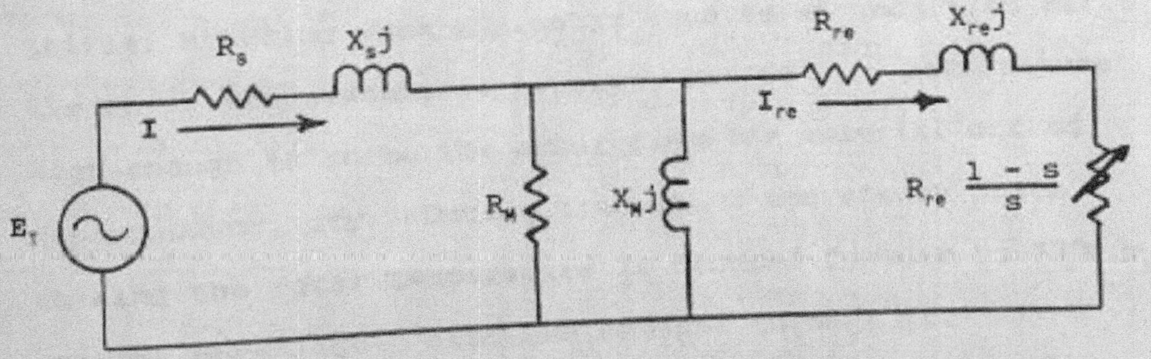

Figure 14 Equivalent circuit for an induction motor during steady state operation.

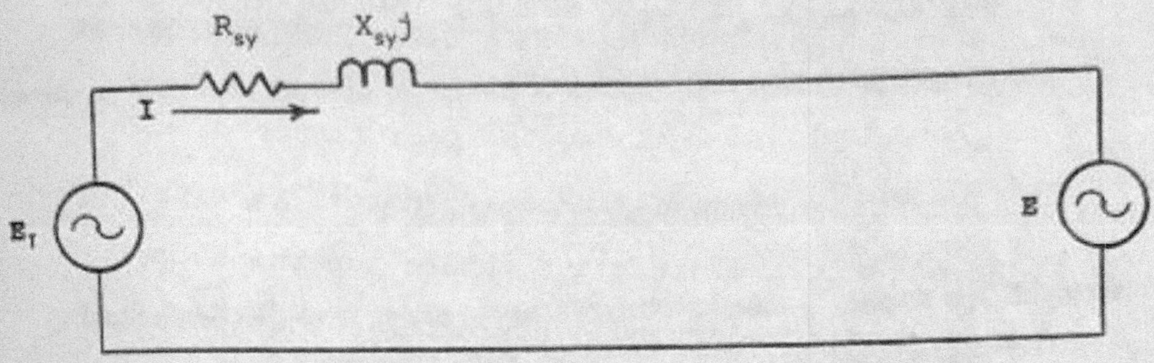

Figure 15 Superconducting synchronous/induction motor steady state phasor equivalent circuit.

This chapter analyzes the same composite squirrel cage rotor as was analyzed in chapters 4.0 and 5.0. It is shown in Figure 1.

Temperature is assumed to be well controlled. During initial starting when the motor runs as an induction motor the rotor temperature is constant at 100°K, a temperature high enough to force the superconductor material out of superconductivity. During transition and steady state running the rotor temperature is dropped further to 77°K by pumping more liquid nitrogen coolant through it.

As in chapter 5.0 it is assumed that the transition to superconductivity occurs suddenly in all superconductors at the same time and that then all of the superconductors stay in superconductivity. This method will not account for unwanted trapped rotor magnetic fluxes.

As in chapters 4.0 and 5.0 these circuits do not account for magnetic flux saturation of the laminations.

6.1 Initial Starting

The induction motor equivalent circuit shown in Figure 14 is used to determine the motor's initial starting performance. This circuit provides accurate estimates of induction motor performance for loads in the range from no load to breakdown, the range in which the superconducting synchronous/induction motor makes the transition to superconductivity.

The equivalent circuit of Figure 14 predicts stator phase peak currents directly and it predicts reflected to the stator values of rotor bar current. Actual rotor bar peak currents are calculated by multiplying the equivalent circuit rotor current by the ratio of stator conductors to rotor bars.

Rotor bar peak current densities and flux densities are calculated from the peak rotor bar currents.

Immediately before the transition to superconductivity, rotor speed and equivalent circuit rotor back emf and phase angle are calculated. Then, these are used to determine the motor's performance during steady state superconductive operation.

Equations (6-1) to (6-7) were written using Figure 14 to describe the superconducting motor while it is starting in the induction motor mode. The 1.35 factor in equations (6-3) and (6-20) approximates the effect of speed change on friction and windage losses. Peak values are used for all AC currents voltages.

6.1.1 Stator Input Current

$$I = \frac{E_T}{R_s + X_s j + \left(\frac{1}{R_M} - \frac{j}{X_M} + \frac{s}{R_{re} + sX_{re} j}\right)^{-1}} \qquad (6\text{-}1)$$

6.1.2 Rotor Current Reflected to the Stator

$$I_{re} = \frac{I}{1 + \left(\frac{1}{R_M} - \frac{j}{X_M}\right)\left(\frac{R_{re}}{s} + X_{re} j\right)} \qquad (6\text{-}2)$$

6.1.3 Mechanical Power from the Rotor

$$P_{out} = 3|I_{re}|^2 R_{re} \frac{1 - s}{s} - P_{fws}(1 - s)^{1.35} - P_{fl} \qquad (6\text{-}3)$$

6.1.4 Mechanical Torque from the Rotor

$$T_{out} = \frac{P_{out}}{w_s(1 - s)} \qquad (6\text{-}4)$$

6.1.5 Peak Rotor Bar Current

An equation for relating equivalent circuit rotor current I_{re} to actual rotor bar current is given by

Fitzgerald and Kingsley[43].

$$I_{rb} = |I_{re}| \frac{mNk_d k_p}{N_b p} \qquad (6\text{-}5)$$

6.1.6 Rotor Back Emf

The rotor produces a back emf that opposes the input voltage. This voltage appears in the equivalent circuit as the voltage across R_M and $X_M j$. After the transition of the rotor bars to superconductivity this back emf voltage is used to predict steady state superconductive back emf voltages in later sections. The phasor voltage is calculated as:

$$E = I_{re} \frac{R_{re}}{s} + X_{re} j \qquad (6\text{-}6)$$

6.1.7 Efficiency

$$E_{ff} = \frac{2T_{out} w_s (1 - s)}{3Re(E_T I^*)} \qquad (6\text{-}7)$$

6.1.8 Rotor Bar Current and Flux Density

The equations (4-28) and (4-29) are used to determine peak current densities and flux densities as they were in sections 4.2.16 and 4.2.17.2.

6.2 Steady State Superconductive Operation

After transition, during steady state operation, the equivalent circuit shown in Figure 15 is used. Its impedances are taken from the induction motor equivalent circuit. R_{sy} is equal to the stator resistance, R_s, plus a corrective resistance value, R_c, that has the same power loss at induction motor stator full load current as R_M had at full load in the induction motor equivalent circuit. X_{sy} is equal to the sum of the induction motor's X_s and X_{re}. In the sum X_s represents stator leakage reactance as it does in an induction motor. X_{re} accounts for the slight reactance that still exists in the rotor and for some of the stator reactance. The use of X_{re} probably produces a conservatively oversized X_{sy}. The reason for the value of X_{sy} being so much lower than that of the usual X_{sy}'s seen in synchronous machines is the trapped flux in the rotor. The trapped flux effect of the superconducting rotor bars will not allow the armature reaction that occurs in ordinary synchronous machines. Figure 15 has two voltage sources. $\mathbf{E}_T$ represents a line to neutral input voltage. $\mathbf{E}$ represents a variable phase angle rotor back emf. The rotor back emf magnitude is derived in section 6.3.1.

Equations (6-8) to (6-12) were written using Figure 15 to describe the superconducting synchronous/induction motor while it is steady state operation in the superconductive mode.

6.2.1 Stator Input Current

$$I = \frac{E_T - |\mathbf{E}(t_o)|\angle\delta_p}{R_{sy} + X_{sy}j} \qquad (6\text{-}8)$$

6.2.2 Mechanical Power from the Rotor

$$P_{out} = 1.5\mathrm{Re}[(|\mathbf{E}(to)|\angle\delta_p)\mathbf{I}^*] - P_{fws} - P_{fl} \qquad (6\text{-}9)$$

6.2.3 Mechanical Torque from the Rotor

$$T_{out} = \frac{P_{out}}{W_s} \qquad (6\text{-}10)$$

6.2.4 Total Input Power, Efficiency, and Power Factor

$$P_{in} = 1.5\mathrm{Re}(\mathbf{E}_T\mathbf{I}^*) \qquad (6\text{-}11)$$

$$E_{ff} = \frac{P_{out}}{P_{in}} \qquad (6\text{-}12)$$

Power factor is the cosine of the angle between $\mathbf{E}_t$ and $\mathbf{I}$.

6.2.5 Rotor Bar Peak Current and Flux Density

Rotor bar peak current and flux density can not be determined from this model directly. A calculation method like that of 5.2.1.6 and 5.2.1.7 must be used. Those methods will not be repeated here.

6.3 Applying the Calculation Methods to an Actual Motor

The motor considered is the same one as in section 4.3 and 5.4, a modification of the 15 hp, 220 V rms, three phase, 6 pole induction motor designed by Kuhlmann[41]. See section 4.3 for a description.

A cross-sectional view of the motor is shown in Figure 1.

During initial starting when the motor is in the induction motor mode the rotor is cooled to 100° K to give the copper part of the rotor bars the same resistance as the full size room temperature rotor bars used by Kuhlmann. With the application of more liquid nitrogen coolant the rotor temperature is switched from 100° K to 77° K causing the transition to superconductivity. Later, during steady state operation, the rotor temperature is maintained continuously at 77° K.

6.4 FORTRAN Programs

Two FORTRAN programs have been written.

The first FORTRAN program solves the steady state induction motor equations (6-1) to (6-7) to produce peak current, stator output torque, power factor, stator current, stator output torque, power factor, efficiency, peak rotor bar current density, peak back emf, and back emf angle relative to the input voltage. These are all calculated versus slip.

The second FORTRAN program solves the steady state synchronous motor like equations (6-8) to (6-12) to produce peak stator current, steady state output torque, power factor, and efficiency versus angle between the source and back emf.

6.4.1 FORTRAN Program Describing the Superconducting Synchronous/Induction Motor in the Normal Conduction Mode

```
C INDUCTION MOTOR STEADY STATE OPERATION
C THI PROGRAM DETERMINES SLIP, STATOR CURRENT, TORQUE, EFFICIENCY
C POWER FACTOR, POWER LOSS, HORSEPOWER, ROTOR PEAK CURRENT DENSITY,
C ROTOR BACK EMF ACROSS THE MAGNETIZING IMPEDANCE AND THE PHASE
C ANGLE OF THE ROTOR BACK EMF ACROSS THE MAGNETIZING IMPEDANCE
C PRIOR TO THE TRANSITION TO SUPERCONDUCTIVITY
C 5/11/89
C
C FOR KUHLMAN'S EXAMPLE THREE PHASE INDUCTION MOTOR
C 15 HP, 60 HZ, 6 POLES, 220 V AND 39.2 A.
C PAGE 351
C
          REAL IR, II, I, IRB, IRE, IRER, IREI, JR,
     2         KD, KP, H, NB, N
          RS =.13
          RRE =.12
          RM = 140.84
          XS =.396
          XRE = .42
          XM = 9.9
          WS = 125.66371
          ET = 179.629248
          M = 3.
          P = 6.
          PFWS = 232.
          PFL = 112.
          N = 162.0
          NB = 65./6.
          KD = .96
          KP = .985
          AB = (5./6.)*.141*.531*.0254**2
          WRITE(6,*) 'RS =', RS
          WRITE(6,*) 'XS =', XS
          WRITE(6,*) 'RRE =', RRE
          WRITE(6,*) 'XRE =', XRE
          WRITE(6,*) 'RM =', RM
          WRITE(6,*) 'XM =', XM
          WRITE(6,*) 'M =', M
          WRITE(6,*) ' S(PU)              I(PU)            TOUT(PU)
     2       EFFICIENCY     POWER FACTOR              PLOSS(WATTS)   HP
     3           JR(PU)                EA1(PU)          DP(PU)'
C CALCULATING TORQUES
          DO 100 S = .000001, .18, .03
            A = RS/RM + XS/XM + 1 + (RS*RRE/S + XS*XRE)/
     2          ((RRE/S)**2 + XRE**2)
            B = -RS/XM + XS/RM - (RS*XRE - XS*RRE/S)/
     2          ((RRE/S)**2 + XRE**2)
            C = 1/RM + RRE/S/((RRE/S)**2 + XRE**2)
            D = -1/XM - XRE/((RRE/S)**2 + XRE**2)
            IR = ET/(A**2 + B**2)*(C*A + D*B)
            II = ET/(A**2 + B**2)*(A*D - C*B)
            I = (IR**2 + II**2)**.5
            PF = IR/I
            E = 1 + RRE/S/RM + XRE/XM
            F = XRE/RM - RRE/S/XM
            IRER = (IR*E + II*F)/(E**2 + F**2)
            IREI = (II*E - IR*F)/(E**2 + F**2)
            IRE = (IRER**2 + IREI**2)**.5
            EA1R = IRER*RRE/S - IREI*XRE
            EA1I = IREI*RRE/S + IRER*XRE
            EA1 = (EA1R**2 + EA1I**2)**.5
            DP = ATAN(EA1I/EA1R)
            IF((EA1R.LT.0).AND.(EA1I.LT.0))THEN
              DP =DP+3.1415927
            ENDIF
            IF((EA1R.LT.0).AND.(EA1I.GE.0))THEN
```

```
            DP =DP*3.1415927
          ENDIF
          IRB = IRE*M*N*KD*KP/NB/P
          JR = IRB/AB
          PIN = M/2*ET*I*PF
          PE = M/2*RRE*(1.0 - S)/S*IRE**2
          WR = WS*(1.0 - S)
          RPM = WR*9.5492966
          PFW = PFWS*(WR/WS)**1.35
          POUT = PE - PFW - PFL
          EFF = POUT/PIN
          TOUT = POUT/WR
          PLOSS = PIN - POUT
          HP = POUT/746
          PUTOUT = TOUT/89.1
          PURPM = RPM/1200
          PUI = I/29.4/1.414214
          PUJR= JR/8.77E6
              PUEA1 = EA1/179.6
              PUDP = DP/.395
          WRITE(6,*) ' '
          WRITE(6,*) S,PUI,PUTOUT,EFF,PF,PLOSS,HP,PUJR,
     2                  PUEA1, PUDP,IRB
100   CONTINUE
      END
```

6.4.2 FORTRAN Program Describing the Superconducting Synchronous/Induction Motor in the Steady State Superconducting Mode

```
C DETERMINING THE CHARACTERISTICS OF THE 15 HP SUPERCONDUCTING
C INDUCTION MOTOR AFTER IT HAS REACHED SYNCHRONOUS SPEED
C AND STOPPED OSCILLATING
C 3/30/89
C
      REAL  I, IETI, IETR, IEOR, IEOI, IR, II, M
      RS = .13
      RC = .06
      XS = .396
      XRE = .42
      EO = 163.16696
      WS = 125.66371
      ET = 179.6292478
      PFWS = 232.
      PFL = 112.
      M = 3.
C
      RSY = RS + RC
      XSY = XS + XRE
             WRITE(6,*) ' DP(RAD)          I(PU)            TOUT(PU)         EFFICI
ENCY
     Z     POWER FACTOR          PLOSS(WATTS)    HP'
      DO 100 DP = 0.0, -.36, -.34
      EOR = EO*COS(DP)
      EOI = EO*SIN(DP)
      IETR = ET*RSY/(RSY**2+XSY**2)
      IETI = -ET*XSY/(RSY**2-XSY**2)
      IEOR = -(RSY*EO*COS(DP)+XSY*EO*SIN(DP))/(RSY**2+XSY**2)
      IEOI = -(-XSY*EO*COS(DP)+RSY*EO*SIN(DP))/(RSY**2+XSY**2)
      IR = IEOR+IETR
      II = IEOI+IETI
      I = (IR**2+II**2)**.5
      PF = IR/I
      PE = M/2*(EOR*IR+EOI*II)
      PIN = M/2*(ET*IR)
      PFW = PFWS
      POUT = PE - PFW - PFL
      EFF = POUT/PIN
      PLOSS = PIN - POUT
      TOUT = POUT/WS
      PUI = I/29.4/1.414214
      PUTOUT = TOUT/89.1
      HP = POUT/746
      WRITE(6,*) ' '
      WRITE(6,*) DP,PUI,PUTOUT,EFF,PF,PLOSS,HP
 100  CONTINUE
      END
```

6.5 Graphed Program Results

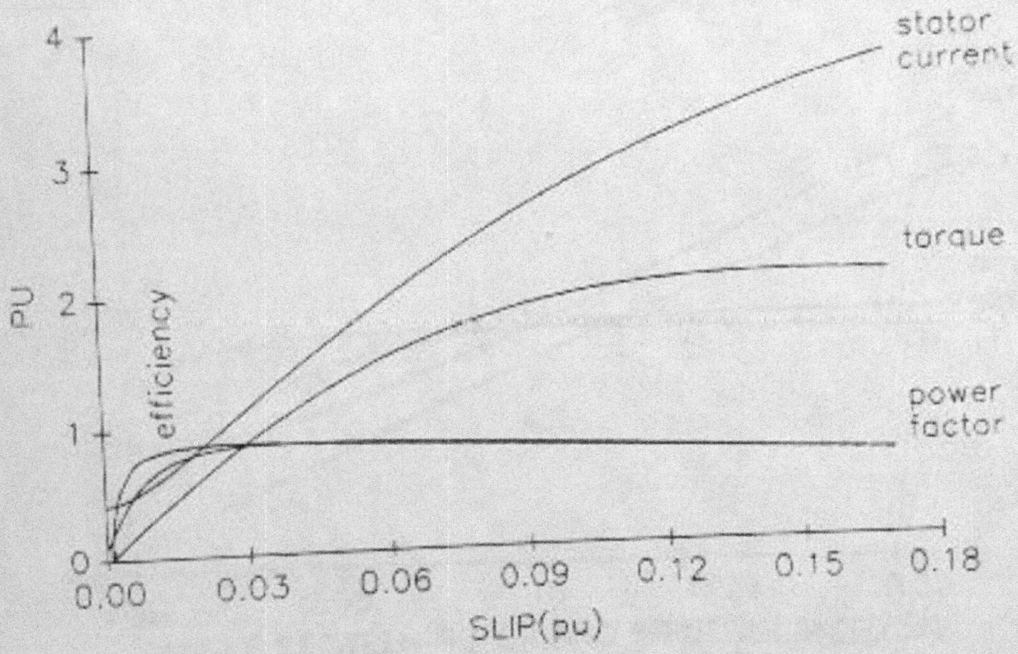

Figure 16 Per unit stator current, torque, power factor, efficiency, and rotor back emf versus slip before superconductivity during normal induction motor operation. Transition to superconductivity occurs when the peak rotor current density is 8.77 x 10^6 A/m^2 by dropping the rotor's temperature. Per unit bases are given in 4.3.2.

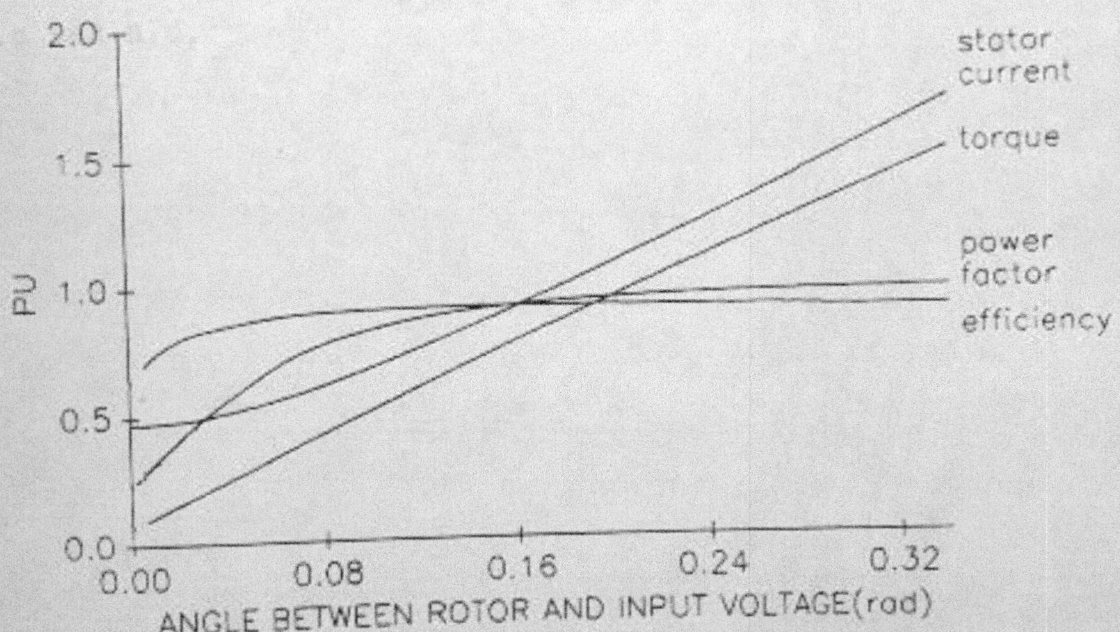

Figure 17 Per unit stator current, torque, power factor, and efficiency versus angle between the rotor back emf and input voltage during steady state superconducting synchronous operation. Per unit bases are given in 4.3.2.

6.6 Observations and Conclusions

The equivalent circuit methods make predictions practically identical to those made with the motor analysis methods of chapters 4.0 and 5.0. The similarity of results tends to validate the methods of this chapter and chapters 4.0 and 5.0.

6.7 Program Output Data

6.7.1 Superconducting Synchronous/Induction Motor in the Normal Conduction Mode

```
$ EX
RUN MOTOR1.EXE

RS =  0.1300000
XS =  0.3960000
RRE =  0.1200000
XRE =  0.4200000
RM =   140.8400
XM =   9.900000
M =   3.000000
 S(PU)            I(PU)              TOUT(PU)           EFFICIENCY       POWER FACTOR
  PLOSS(WATTS)    HP                 JR(PU)             EA1(PU)           DP(PU)

 1.0000000E-06    0.4202540      -3.0690273E-02  -0.9113830         8.0083735E-02
  720.6671      -0.4606266        2.8799603E-05   0.9607943         0.1043200
 1.0166838E-02

 1.5001000E-02    0.6929733        0.4484746        0.8478926         0.7514032
  887.3003        6.630135         0.4236217        0.9431854        -0.3792462
  149.5117

 3.0001000E-02     1.130397        0.8744894        0.8746981         0.8574198
  1360.545        12.73136         0.8231959        0.9204338        -0.8167571
  290.6047

 4.5001000E-02     1.565670         1.233397        0.8644489         0.8698071
  2068.031        17.67887          1.191428        0.8941508        -1.198640
  420.5981

 6.0001001E-02     1.967973         1.520774        0.8452225         0.8589317
  2931.005        21.45561          1.524125        0.8659192        -1.520360
  538.0467

 7.5001001E-02     2.329459         1.739611        0.8228943         0.8389791
  3877.658        24.15139          1.820249        0.8371010        -1.781754
  642.5845

 9.0001002E-02     2.649284         1.897476        0.7994362         0.8148199
  4850.355        25.91588          2.081008        0.8087497        -1.985917
  734.6378

 0.1050010         2.929669         2.003971        0.7758798         0.7888049
  5807.475        26.91923          2.309010        0.7816016        -2.138010
  815.1270

 0.1200010         3.174218         2.068925        0.7520252         0.7622799
  6721.859        27.32597          2.507561        0.7561163        -2.244272
  885.2195

 0.1350010         3.387019         2.101348        0.7286806         0.7360841
  7577.832        27.28113          2.680171        0.7325355        -2.311237
  946.1541

 0.1500010         3.572140         2.108961        0.7057574         0.7106568
  8368.061        26.90517          2.830241        0.7109438        -2.345239
  999.1318

 0.1650010         3.733369         2.098093        0.6833164         0.6863483
  9090.795        26.29416          2.960897        0.6913194        -2.352108
  1045.256
```

6.7.2 Superconducting Synchronous/Induction Motor in the Steady State Superconducting Mode

[illegible](RAD)	I(PU)	TOUT(PU)	EFFICIENCY	POWER FACTOR
PLOSS(WATTS)	HP			
0.0000000E+00	0.4725766	6.6678807E-02	0.6218329	0.2267789
154.0309	1.000775			
-2.5000000E-02	0.4882866	0.1804473	0.8140644	0.4537050
461.4580	2.708314			
-5.0000001E-02	0.5326343	0.2934805	0.8716699	0.6317639
483.7747	4.404818			
-7.5000003E-02	0.5992718	0.4057077	0.8971192	0.7542164
520.9370	6.089224			
-0.1000000	0.6816592	0.5170588	0.9099483	0.8331299
572.9312	7.760481			
-0.1250000	0.7747458	0.6274642	0.9165418	0.8831496
639.7251	9.417544			
-0.1500000	0.8750755	0.7368549	0.9196041	0.9151503
721.2783	11.05938			
-0.1750000	0.9803743	0.8451624	0.9204769	0.9360352
817.5391	12.68496			
-0.2000000	1.089146	0.9523196	0.9199011	0.9499747
928.4434	14.29326			
-0.2250000	1.200389	1.058259	0.9183183	0.9594743
1053.929	15.88329			
-0.2500000	1.313414	1.162914	0.9160081	0.9660582
1193.914	17.45405			
-0.2750000	1.427737	1.266220	0.9131564	0.9706724
1348.308	19.00456			
-0.3000000	1.543004	1.368112	0.9098905	0.9739177
1517.018	20.53385			
-0.3250000	1.658955	1.468526	0.9063008	0.9761837
1699.938	22.04098			
-0.3500001	1.775391	1.567400	0.9024531	0.9777281
1896.953	23.52495			

7.0 TWO AXIS THEORY METHOD FOR DETERMINING SUPERCONDUCTING SYNCHRONOUS/INDUCTION MOTOR PERFORMANCE

The two axis theory method of the type described by Brown and Hamilton[44] is used to determine the performance of superconducting synchronous/induction motors.

The two axis method uses systems of nonlinear differential equations to describe motor performance. Similar differential equations are used to describe superconducting synchronous/induction motors during normal conductive operation, steady state superconductive operation, and transition to superconductivity.

This chapter analyzes the same composite squirrel cage rotor as was analyzed in chapters 4.0, 5.0, and 6.0. It is shown in Figure 1.

Temperature is assumed to be well controlled. During initial starting when the motor runs as an induction motor the rotor temperature is constant at 100°K, a temperature high enough to force the superconductor material out of superconductivity. During transition and steady state running the rotor temperature is dropped further to 77°K by pumping more liquid nitrogen coolant through it.

As in chapters 5.0 and 6.0 it is assumed that the transition to superconductivity occurs suddenly in all superconductors at the same time and then all of the superconductors stay in superconductivity.

As in chapters 4.0, 5.0, and 6.0 the two axis method does not account for magnetic flux saturation of the laminations and will not account for unwanted trapped rotor magnetic fluxes.

Rotor bar peak current and flux density can not be directly determined with the two axis method. A calculation method like that of 5.2.1.6 and 5.2.1.7 must be used. Those methods will not be repeated here.

7.1 Transition to Superconductivity

The performance of the motor during transition from normal conduction to superconduction is difficult to accurately predict.

Since different rotor bars are experiencing different current and flux densities at the same time it seems likely that some rotor bars will become superconductive while others will not. The superconductive rotor bars will stop heating resistively, tending to keep them superconductive. The normally conducting rotor bars will continue to heat resistively, tending to keep them in normal conduction. The superconductive rotor bars will trap flux if they form into complete superconductive circuits with other superconductive rotor bars. The trapped flux will attempt to accelerate the rotor, but during acceleration excessive currents or flux densities in the superconductive rotor bars may drive them out of superconductivity again. Current, torque, and speed oscillations will occur as rotor bars go in and out of superconductivity.

In analysis this problem has been avoided by using the simplification that all superconductor material goes into superconductivity at the same time and then stays superconductive. Details of the cooling system are not considered here. This analysis looks only at the electrical part of the problem. It is simply assumed that coolant suddenly drops the temperature of the superconductor material to a constant 77° K, causing superconductivity to suddenly occur throughout the superconductor material.

In an actual superconducting synchronous/induction motor there are two other effects that might help speed a superconducting rotor up to synchronous operating speed. First, with the use of superconductor material of small enough current and flux density capacity, it is felt that undesirable local trapped fluxes will be wiped out by superconductor current density or flux density saturation as the rotor revolves relative to stator's rotating field. Second, the results in chapter 4.3.4 show that as the rotor bar resistance decreases the motor output torque increases. This would accelerate the rotor as more and more sections of the rotor bars go into superconductivity.

It is believed that more practical superconductor materials need to be developed and actual experiments need to be done to accurately determine transition performance.

7.1 Two Axis Theory Equations

The induction motor's two axis equivalent is shown in Figure 18. This is the same figure as given by Brown and Hamilton[44].

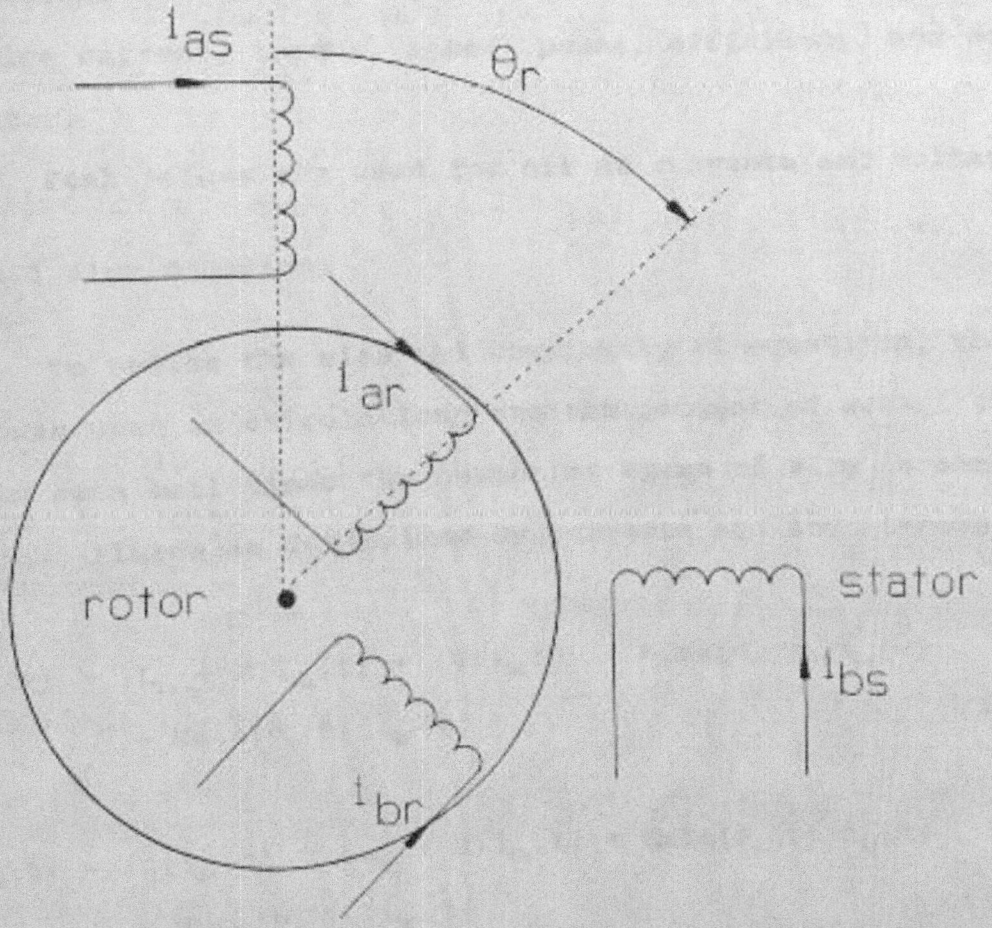

Figure 18 Two axis equivalent for a superconducting synchronous/induction motor.

The two axis method of Figure 18 determines stator and rotor fluxes through each coil in each phase. Fluxes are calculated two ways. First, they are calculated using leakage and mutual inductances times currents. Second, they are calculated as the integral of their voltage rises minus voltage drops. The equations describing each flux are equated. Then these equations are combined with equations involving motor load, torque, and inertia. Finally, using numerical methods, the equations are solved. The solution yields current, torque, speed, power, efficiency, and power factor.

Peak values are used for all AC currents and voltages.

7.1.1 Flux Equations

To reduce the size and complexity of equations, the fluxes used in calculations are the product of actual flux from each coil times the number of turns of wire in each coil. Fluxes as determined by currents and inductances:

$$\phi_{as}(t) = (L_{ls} + M)i_{as}(t) + (0)i_{bs}(t) + M\cos[\theta_r(t)]i_{ar}(t) - M\sin[\theta_r(t)]i_{br}(t) \quad (7\text{-}1)$$

$$\phi_{bs}(t) = (0)i_{as}(t) + (L_{ls} + M)i_{bs}(t) + M\sin[\theta_r(t)]i_{ar}(t) + M\cos[\theta_r(t)]i_{br}(t) \quad (7\text{-}2)$$

$$\phi_{ar}(t) = M\cos[\theta_r(t)]i_{as}(t) + M\sin[\theta_r(t)]i_{bs}(t) + (L_{lr} + M)i_{ar}(t) + (0)i_{br}(t) \quad (7\text{-}3)$$

$$\phi_{br}(t) = -M\sin[\theta_r(t)]i_{as}(t) + M\cos[\theta_r(t)]i_{bs}(t) + (0)i_{ar}(t) + (L_{lr} + M)i_{br}(t) \quad (7\text{-}4)$$

Fluxes as determined by voltage rises minus drops:

$$\phi_{as}(t) = \int [v_{as}(t) - r_{as}i_{as}(t)]dt \quad (7\text{-}5)$$

$$\phi_{bs}(t) = \int [v_{bs}(t) - r_{bs}i_{bs}(t)]dt \quad (7\text{-}6)$$

$$\phi_{ar}(t) = \int [- r_{ar}i_{ar}(t)]dt \quad (7\text{-}7)$$

$$\phi_{br}(t) = \int [- r_{br}i_{br}(t)]dt \quad (7\text{-}8)$$

7.1.2 Mechanical Power from the Rotor

Mechanical power from the rotor is the input power minus the losses. In order to be closer to the usual two axis method the fixed losses incorporated in previous models in chapters 4.0, 5.0, and 6.0 will not be used. Since those losses are only a small part of the total losses the effect of the omission is small. The 1.35 factor approximates the effect of speed change on friction and windage losses.

$$P_{out}(t) = v_{as}(t)i_{as}(t) + v_{bs}(t)i_{bs}(t) - i_{as}(t)^2 r_{as} - i_{bs}(t)^2 r_{as} - i_{ar}(t)^2 r_{ar} - i_{br}(t)^2 r_{br} - P_{fws}[w_r(t)/w_s(t)]^{1.35} \quad (7\text{-}9)$$

7.1.3 Mechanical Torque from the Rotor

$$T_{out}(t) = \frac{P_{out}(t)}{w_r(t)} \quad (7\text{-}10)$$

7.1.4 Efficiency

$$P_{in}(t) = v_{as}(t)i_{as}(t) + v_{bs}(t)i_{bs}(t) \quad (7\text{-}11)$$

$$E_{ff}(t) = \frac{P_{out}(t)}{P_{in}(t)} \quad (7\text{-}12)$$

7.1.5 Power Factor

Power factor is calculated as the ratio of average input power to the product of two times the rms current and voltage. Integration versus time is used to determine the rms current and average power as they are not defined as simple sinusoidal functions or magnitudes, but are solutions of numerically solved differential equations.

$$P_{av}(t) = \frac{\int_0^T P_{in}(t)dt}{T} \quad (7\text{-}13)$$

$$I_{rms}(t) = \left[\frac{\int_0^T i_{as}(t)^2 \, dt}{T} \right]^{.5} \tag{7-14}$$

$$PF(t) = \frac{P_{av}(t)}{1.414 I_{rms}(t) V_{as}} \tag{7-15}$$

7.1.6 Equations and Quantities Particular to the State of the Superconducting Synchronous/Induction Motor

7.1.6.1 Rotor Resistance. The most obvious quantity that changes as the superconducting synchronous/induction motor makes the transition from normal conduction to superconductivity is the rotor resistance. It changes from a finite value to zero.

7.1.6.2 Rotor Coil Flux. When an induction motor is in steady state operation the fluxes through the rotor coils vary as a function of time.

When the superconducting synchronous/induction motor is making the transition from normal conduction to superconduction the fluxes through the rotor axes become fixed. In transition computations the flux values used are the last ones that were computed when the motor was operating as an induction motor.

When the superconducting synchronous/induction motor is in steady state superconducting synchronous motor operation the fluxes through the rotor axes are constant. The flux values were computed when the motor was in steady state induction motor operation.

7.1.6.3 Angular Position and Speed of the Rotor. When the superconducting synchronous/induction motor is in steady state induction motor operation the angular speed, $w_r(t)$, is constant. The equation for rotor position simplifies to:

$$\theta_r(t) = w_s(1 - s)t \qquad (7\text{-}16)$$

The initial angular position is assumed to be 0 for computations.

When the superconducting synchronous/induction motor is in steady state superconducting synchronous motor operation the angular speed of the rotor is constant at w_s. The equation for rotor position simplifies to:

$$\theta_r(t) = w_s t + \theta_r(t_o) \qquad (7\text{-}17)$$

When the superconducting synchronous/induction motor is making the transition from normal conduction to superconduction the rotor's position is calculated as the integral of its angular speed versus time. The rotor's

$$\theta_r(t) = \int w_r(t)dt + \theta_r(t_o) \tag{7-18}$$

In using this equation in the computer programs the initial angle is not too important. The motor tends to go through similar oscillations soon after starting regardless of the angle. The initial value of $w_r(t)$ is selected to be near the value seen during steady state induction motor operation.

7.1.6.4 <u>Rotor Speed</u> for the case of the transition to superconductivity is calculated as:

$$w_r(t) = \int \frac{T_{out}(t) - T_l}{J} dt + w_r(t_o) \tag{7-19}$$

7.1.6.5 <u>Average Rotor Speed and Torque</u> are calculated for the case of transition to superconductivity as:

$$T_{av}(t) = \frac{\int_0^T T_{out}(t)dt}{T} \tag{7-20}$$

$$w_{rav}(t) = \frac{\int_0^T w_r(t)dt}{T} \tag{7-21}$$

7.2 Applying the Two Axis Theory Method to an Actual Motor

The motor considered is the same one as in section 4.3, 5.4, and 6.4. It is a modification of the 15 hp, 220 V rms, three phase, 6 pole induction motor designed by Kuhlmann[41]. See section 4.3 for a description.

A cross-sectional view of the motor is shown in Figure 1.

During initial starting when the motor is in the induction motor mode the rotor is cooled to 100° K to give the copper part of the rotor bars the same resistance as the full size room temperature rotor bars used by Kuhlmann. With the application of more liquid nitrogen coolant the rotor temperature is switched from 100° K to 77° K causing the transition to superconductivity. Later, during steady state operation, the rotor temperature is maintained continuously at 77° K.

The impedance and voltage values used in chapters 4.0, 5.0, and 6.0 are converted to those used in the two axis method. The values are converted so that the two axis method predicts the same induction motor input power, output power, and input current.

The lumped impedance values previously used apply to an equivalent circuit that describes one Y leg of the 3 phase motor being studied. That way three equivalent circuits are required to completely describe the three phase motor.

The two axis theory of Hamilton and Brown[44] describes a 2 phase induction motor. They have shown that their two axis description can also be reduced to an equivalent circuit. Their equivalent circuit represents one of the two phases of their two phase motor. With their methods two equivalent circuits are required to completely describe their two phase motor.

The conversion of lumped impedance equivalent circuit values will be explained intuitively. After that the results of that conversion will be checked more rigorously.

Consider an imaginary simple 3 coil, 2 pole three phase motor. Suppose each phases' winding contains N turns. Remove the coils from the stator. Cut new stator slots and use the same coil wire to wind a two phase winding onto the stator through the new, two phase, slots. Logically, there should be 1.5 times as many turns on each of the stator's new 2 stator coils. Therefore the resistance of each of the coils should be 1.5 times that seen formerly with the 3 phase coils. Also, the inductive leakage impedance of each of the coils should be 1.5 times that seen formerly with the 3 phase coils.

The times 1.5 factor is applied to the stator portion of the two phase equivalent circuit. Although it is not so obvious the 1.5 times factor should be also be applied to the rotor and magnetizing impedance portions of the equivalent circuit.

Our newly created two phase motor is to be equal in power capacity and operation to the three phase motor that it was created from. The wire that was used is rated to a certain current value. It seems logical to assume that the wire should carry the same current values as it previously carried in the three phase motor it was formerly in. Also the new two phase motor should use the same amount of apparent power as it did as a three phase motor for the same input current.

$$P_{3phase} = 3V_{3phase}I \tag{7-22}$$

$$P_{2phase} = 2V_{2phase}I \tag{7-23}$$

Equating these the voltage to be applied to our two phase equivalent circuit is 1.5 times that applied to our three phase equivalent circuit.

The two equivalent circuits are given below.

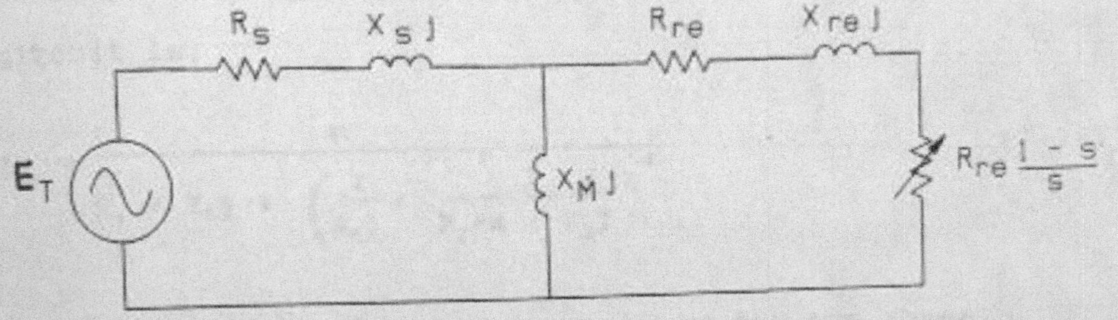

Figure 19 Equivalent circuit applicable to one Y leg of a 3 phase induction motor.

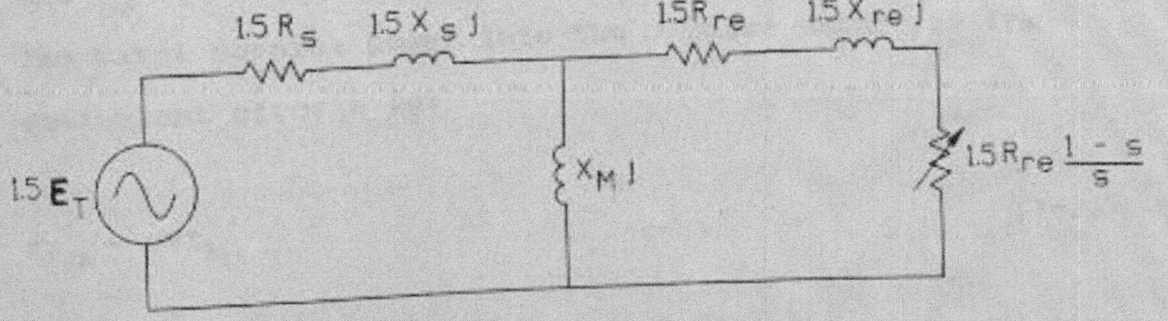

Figure 20 Equivalent circuit applicable to one phase of a 2 phase induction motor. The voltage and impedance values have been multiplied by 1.5 times the values used in the one Y leg equivalent circuit.

The transformations can be checked by comparing the calculated the input power, output power, and phasor input current. The input current to the three phase equivalent circuit is:

$$I_{3p} = \frac{V}{R_1 + X_1 j + \left(\frac{1}{X_M j} + \frac{1}{R_2/s + X_2 j}\right)^{-1}} \quad (7\text{-}24)$$

The input current to the one phase of the two phase equivalent circuit is:

$$I_{2p} = \frac{1.5V}{1.5R_1 + 1.5X_1 j + \left(\frac{1}{1.5X_M j} + \frac{1}{1.5R_2/s + 1.5X_2 j}\right)^{-1}} \quad (7\text{-}25)$$

By inspection it is seen that the 1.5 factor cancels out of the I_{2p} equation, so the currents are the same.
The total complex power into the 3 phase motor via its equivalent circuit is:

$$P_{in3p} = 3VI_{3p}^{*} \quad (7\text{-}26)$$

The total complex power into the 2 phase motor via its equivalent circuit is:

$$P_{in2p} = 2(1.5V)I_{2p}^{*} \quad (7\text{-}27)$$

Since the currents are the same by inspection we can see that the input powers are the same.

The total output power from the 3 phase motor's equivalent circuit is:

$$P_{out3p} = 3 \left| \frac{X_M j}{X_M j + R_2/s + X_2 j} I_{3p} \right|^2 R_2 \frac{1 - s}{s} \qquad (7\text{-}28)$$

The total output power from the 2 phase motor's equivalent circuit is:

$$P_{out2p} = 2 \left| \frac{1.5X_M j}{1.5X_M j + 1.5R_2/s + 1.5X_2 j} I_{2p} \right|^2 1.5R_2 \frac{1 - s}{s} \qquad (7\text{-}29)$$

Since the currents are the same by inspection we can see that the output powers are the same.

Therefore our two phase equivalent circuit will produce the same steady state values as the previously used three phase equivalent circuit.

As was mentioned previously, Brown and Hamilton[44] have shown that their equivalent circuit can be derived from their two axis theory. Therefore the values that were converted from the three axis equivalent circuit can then be applied to their two axis theory.

7.3 CSMP Programs

Three CSMP programs have been written.

Each program assumes initial currents are zero. Initial speeds were fixed for the steady state normal and superconducting operations at a slip speed or synchronous speed. For the transition program the initial speed was assumed to be close to normal slip speed.

The first CSMP program solves the equations for normal induction motor steady state operation. The program produces stator current, output torque, power factor, and efficiency versus slip. The program's output asymptotically approached a reasonable and steady state output after several fractions of a second of motor running.

The second CSMP program solves the equations for steady state superconducting synchronous operation. The program produces stator current, output torque, power factor, and efficiency versus rotor initial angular position. The program's output asymptotically approached a reasonable and steady state output after several fractions of a second of motor running.

The third CSMP program solves the equations for operation during and immediately after the transition to superconductivity. The program produces instantaneous stator current, output torque, and rotor speed versus time. The program's output indicates the motor's operation is unstable.

7.3.1 CSMP Program Describing the Superconducting Synchronous/Induction Motor in the Normal Conduction Mode

```
* STEADY STATE INDUCTION MOTOR
* 11/14/90
* KUHLMANN'S 15 HP, 220 V, 3 PHASE INDUCTION MOTOR
*
CONSTANTS LLR2 = .00167, LLS2 = .00155, M2 = .0394, ...
          RR2 = .18, RS2 = .225, WS = 125.66370, ...
          S = .0375, VM = 269.56, ...
          PFWS = 232.0, ...
          IAS0 = 0, IBS0 = 0, IAR0 = 0, IBR0 = 0
*
       THR = 3*WS*(1 - S)*TIME
*
*      A STATOR COIL FLUX
       IAS = (1/(LLS2 + M2))*(-VM/(3*WS)*COS(3*WS*TIME) ...
             - RS2*INTGRL(IAS0, IAS) - M2*COS(THR)*IAR ...
             + M2*SIN(THR)*IBR)
*
*      B STATOR COIL FLUX
       IBS = (1/(LLS2 + M2))*(-VM/(3*WS)*COS(3*WS*TIME - ...
             1.570796) - RS2*INTGRL(IBS0, IBS) - ...
             M2*SIN(THR)*IAR  - M2*COS(THR)*IBR)
*
*      A ROTOR COIL FLUX
       IAR = (1/(LLR2 + M2))*(-RR2*INTGRL(IAR0, IAR) ...
             - M2*COS(THR)*IAS - M2*SIN(THR)*IBS)
*
*      B ROTOR COIL FLUX
       IBR = (1/(LLR2 + M2))*(-RR2*INTGRL(IBR0, IBR) ...
             + M2*SIN(THR)*IAS - M2*COS(THR)*IBS)
*
       VAS = VM*SIN(3*WS*TIME)
       VBS = VM*SIN(3*WS*TIME - 1.570796)
       PIN = VAS*IAS + VBS*IBS
       PFW = PFWS*(1 - S)**1.35
       POUT = PIN - IAS**2*RS2 - IBS**2*RS2 - ...
              IAR**2*RR2 - IBR**2*RR2 - PFW
       EFF = POUT/PIN
       TOUT = POUT/((1 - S)*WS)
       PUTOUT = TOUT/89.1
       WR = (1 - S)*WS
       RMSI1 = IAS**2*STEP(.1)
       RMSI2 = INTGRL(0.0, RMSI1)
       RMSI = (STEP(.10001)*RMSI2/(TIME-.1))**.5
       PUIAS = RMSI/29.4
       PINI = PIN*STEP(.1)
       PINAV1 = INTGRL(0.0, PINI)
       PINAV = STEP(.10001)*PINAV1/(TIME-.1)
       PF = PINAV/(2*(RMSI+.00000001)*VM/1.4142135)
       LAR = M2*COS(THR)*IAS + M2*SIN(THR)*IBS + (LLR2 + M2)*IAR
       LBR = -M2*SIN(THR)*IAS + M2*COS(THR)*IBS + (LLR2 + M2)*IBR
  TIMER FINTIM = .2, OUTDEL = .05, PRDEL = .05
  PRINT S, EFF, PF, PUTOUT, PUIAS, LAR, LBR
  END
  STOP
```

7.3.2 CSMP Program Describing the Superconducting Synchronous/Induction Motor in the Steady State Superconducting Mode

```
* STEADY STATE SUPERCONDUCTING SYNCHRONOUS MOTOR
* 11/14/90
*
* KUHLMANN'S 15 HP, 220 V, 6 POLE, 3 PHASE INDUCTION MOTOR
*
CONSTANTS LLR2 = .00167, LLS2 = .00155, M2 = .0394, ...
          RS2 = .225, WS = 125.66371, ...
          VM = 269.56, PFWS = 232.0, ...
          IAS0 = 0, IBS0 = 0, IAR0 = 0, IBR0 = 0, ...
          LAR0 = -.14454, LBR0 = -.62387, ...
* THR01 = 0, .0175, .035, .07, .105, .14, .175, .21, .245,
*         .28, .315, .35
          THR01 = .21
          THR0 = -THR01 - 1.334
*
       THR = 3*WS*TIME + THR0
*
*      A STATOR COIL FLUX
       IAS = (1/(LLS2 + M2))*(-VM/(3*WS)*COS(3*WS*TIME) ...
             - RS2*INTGRL(IAS0, IAS) - M2*COS(THR)*IAR ...
             + M2*SIN(THR)*IBR)
*
*      B STATOR COIL FLUX
       IBS = (1/(LLS2 + M2))*(-VM/(3*WS)*COS(3*WS*TIME ...
             - 1.570796) - RS2*INTGRL(IBS0, IBS) ...
             - M2*SIN(THR)*IAR - M2*COS(THR)*IBR)
*
*      A ROTOR COIL FLUX
       IAR = (1/(LLR2 + M2))*(LAR0 ...
             - M2*COS(THR)*IAS - M2*SIN(THR)*IBS)
*
*      B ROTOR COIL FLUX
       IBR = (1/(LLR2 + M2))*(LBR0 ...
             + M2*SIN(THR)*IAS - M2*COS(THR)*IBS)
*
       VAS = VM*SIN(3*WS*TIME)
       VBS = VM*SIN(3*WS*TIME - 1.570796)
       PIN = VAS*IAS + VBS*IBS
       PFW = PFWS
       POUT = PIN - IAS**2*RS2 - IBS**2*RS2 - PFW
       EFF = POUT/PIN
       TOUT = POUT/WS
       PUTOUT = TOUT/89.1
       RMSI1 = IAS**2*STEP(.1)
       RMSI2 = INTGRL(0.0, RMSI1)
       RMSI = (STEP(.10001)*RMSI2/(TIME-.1))**.5
       PUIAS = RMSI/29.4
       PIN1 = PIN*STEP(.1)
       PINAV1 = INTGRL(0.0, PIN1)
       PINAV = STEP(.10001)*PINAV1/(TIME - .1)
       PF = PINAV/(2*(RMSI + .00000001)*VM/1.4142135)
TIMER FINTIM = .2, OUTDEL = .05, PRDEL = .05
PRINT THR01, EFF, PF, PUTOUT, PUIAS
END
STOP
```

7.3.3 CSMP Program Describing the Superconducting Synchronous/Induction Motor in Transition from the Normal Conduction Mode to the Superconducting Mode

```
* TRANSITION FROM STEADY STATE INDUCTION MOTOR TO
* SUPERCONDUCTING SYNCHRONOUS MOTOR
* 3/4/91
*
* KUHLMANN'S 15 HP, 220 V, 6 POLE, 3 PHASE INDUCTION MOTOR
*
* TIME OF TRANSITION IS .1, THE PROGRAM IS RUN WITH A .11 TRANSITION
* TIME TO ALLOW LARO AND LBRO TO BE DETERMINED
*
CONSTANTS LLR2 = .00167, LLS2 = .00155, M2 = .0394, ...
          RS2 = .225, WS = 125.66371, ...
          VM = 269.56, PFWS = 232.0, ...
          IAS0 = 0, IBS0 = 0, IARO = 0, IBRO = 0, ...
          LARO = .50303, LBRO = .34837, ...
          WRO = 121.01414, TL = 89.1, J = .5
 *
        RR2 = .18*(1 - STEP(.1))
 *
        THR = 3*INTGRL(1.5, WR)
 *
 *      A STATOR COIL FLUX
        IAS = (1/(LLS2 + M2))*(-VM/(3*WS)*COS(3*WS*TIME) ...
              - RS2*INTGRL(IAS0, IAS) - M2*COS(THR)*IAR ...
              + M2*SIN(THR)*IBR)
 *
 *      B STATOR COIL FLUX
        IBS = (1/(LLS2 + M2))*(-VM/(3*WS)*COS(3*WS*TIME ...
              - 1.570796) - RS2*INTGRL(IBS0, IBS) ...
              - M2*SIN(THR)*IAR - M2*COS(THR)*IBR)
 *
 *      A ROTOR COIL FLUX
        IAR = (1/(LLR2 + M2))*(-RR2*INTGRL(IARO, IAR) ...
              + LARO*STEP(.1) ...
              - M2*COS(THR)*IAS - M2*SIN(THR)*IBS)
 *
 *      B ROTOR COIL FLUX
        IBR = (1/(LLR2 + M2))*(-RR2*INTGRL(IBRO, IBR) ...
              + LBRO*STEP(.1) ...
              + M2*SIN(THR)*IAS - M2*COS(THR)*IBS)
 *
        LAR = M2*COS(THR)*IAS + M2*SIN(THR)*IBS + (LLR2 + M2)*IAR
        LBR = -M2*SIN(THR)*IAS + M2*COS(THR)*IBS + (LLR2 + M2)*IBR
        VAS = VM*SIN(3*WS*TIME)
        VBS = VM*SIN(3*WS*TIME - 1.570796)
        PIN = VAS*IAS + VBS*IBS
        PFW = PFWS*(WR/WS)**1.35
        POUT = PIN - IAS**2*RS2 - IBS**2*RS2 - ...
               IAR**2*RR2 - IBR**2*RR2 - PFW
        EFF = POUT/PIN
        TOUT = POUT/WR
        DWRDT = 1/J*(TOUT - TL)
        WR = INTGRL(WRO, DWRDT)
        PUTOUT = TOUT/89.1
        PUIAST = IAS/29.4/1.41421356
        PURPM = WR/125.6637
        RTIME = TIME - .1
        PUTAV1 = INTGRL(0.0, PUTOUT)
        PUTAV = PUTAV1/(TIME + .000000001)
        RPMAV1 = INTGRL(0.0, PURPM)
        RPMAV = RPMAV1/(TIME + .000000001)
TIMER FINTIM = .7, OUTDEL = .02, PRDEL = .02
*PRTPLT PUTOUT
*PRTPLT PURPM
PRINT RTIME, PUTOUT, PURPM, PUTAV, RPMAV
END
STOP
ENDJOB
```

7.4 Graphed Program Results

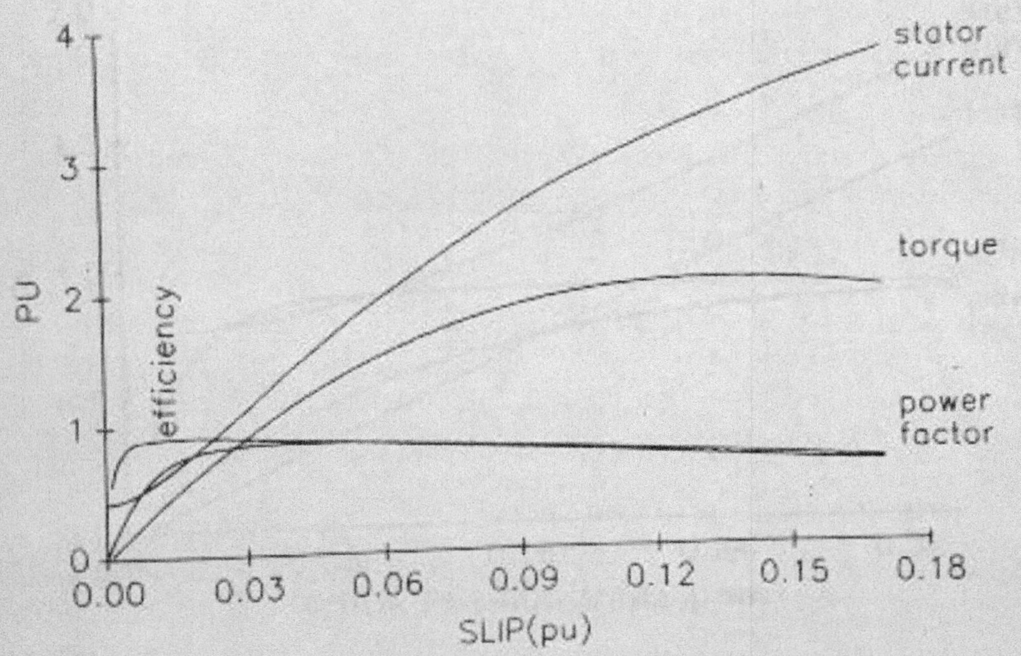

Figure 21 Per unit stator current, torque, power factor, efficiency, and rotor back emf versus slip before superconductivity during normal induction motor operation. Transition to superconductivity occurs when the slip is .037 by dropping the rotor bars' temperature. Per unit bases are given in 4.3.2.

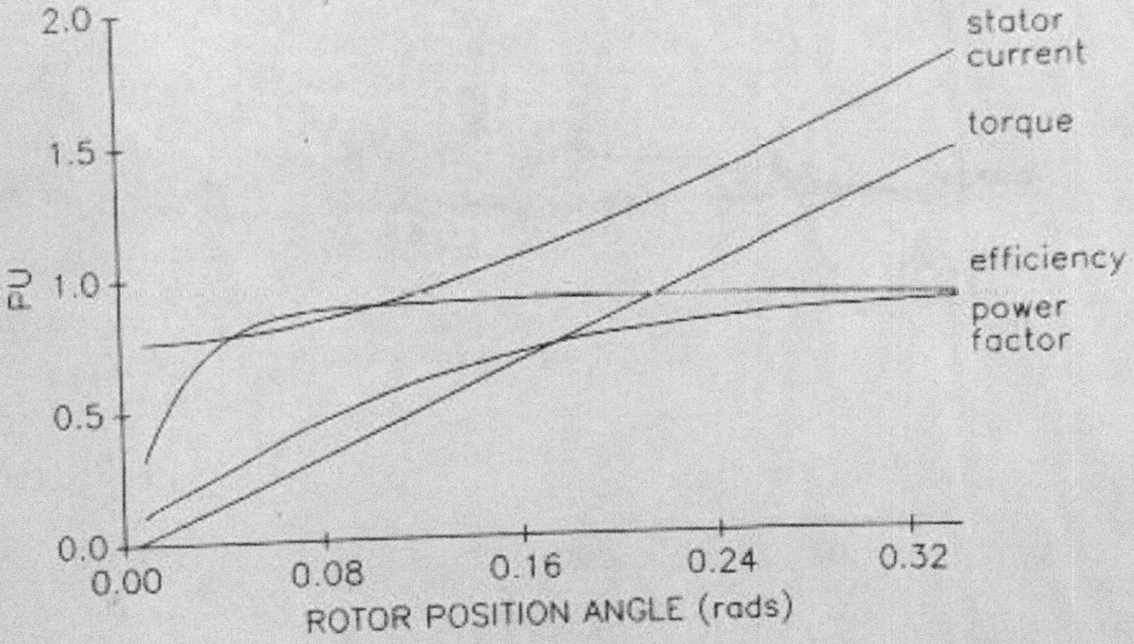

Figure 22 Per unit stator current, torque, power factor, and efficiency versus angle between the rotor initial angle and input voltage during steady state superconducting synchronous operation. Per unit bases are given in 4.3.2.

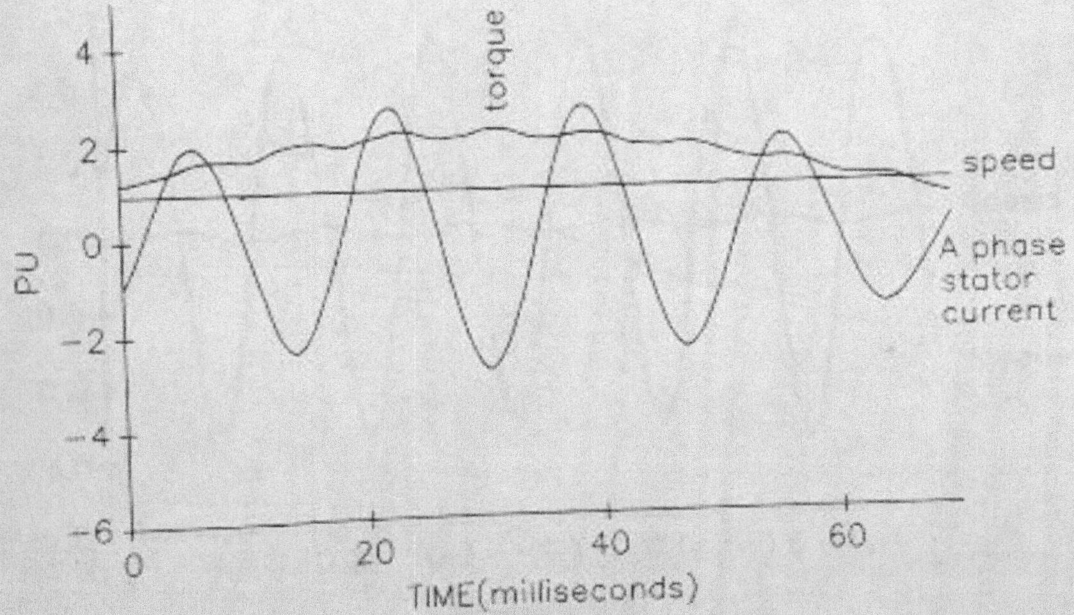

Figure 23 Per unit current, torque, and speed versus time during starting immediately after the transition to superconductivity. Per unit bases are given in 4.3.2.

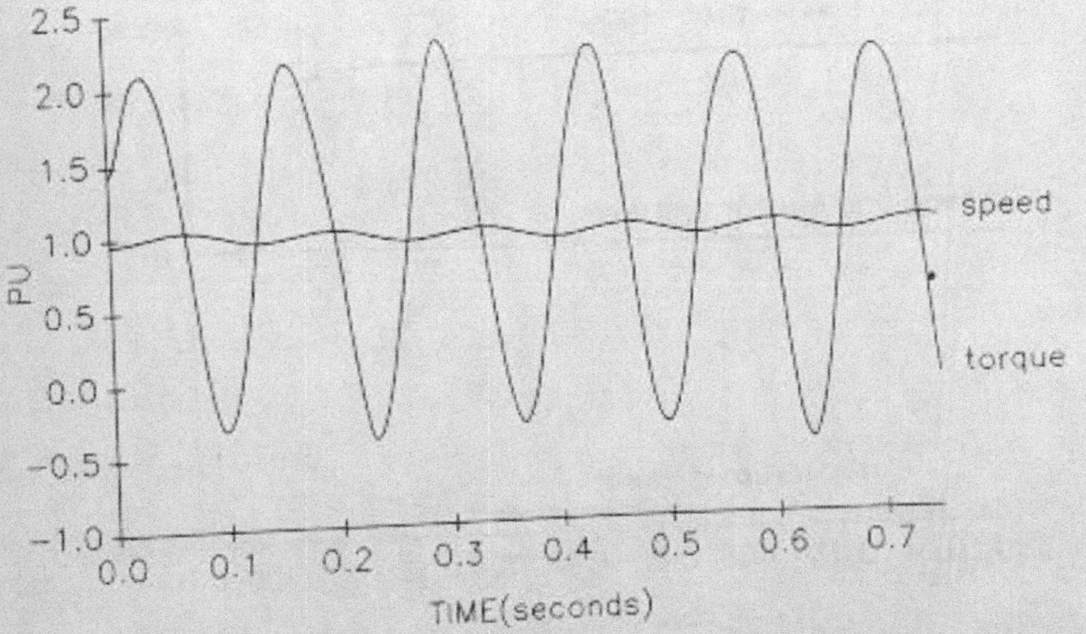

Figure 24 Per unit torque and speed versus time during starting immediately after the transition to superconductivity. Per unit bases are given in 4.3.2.

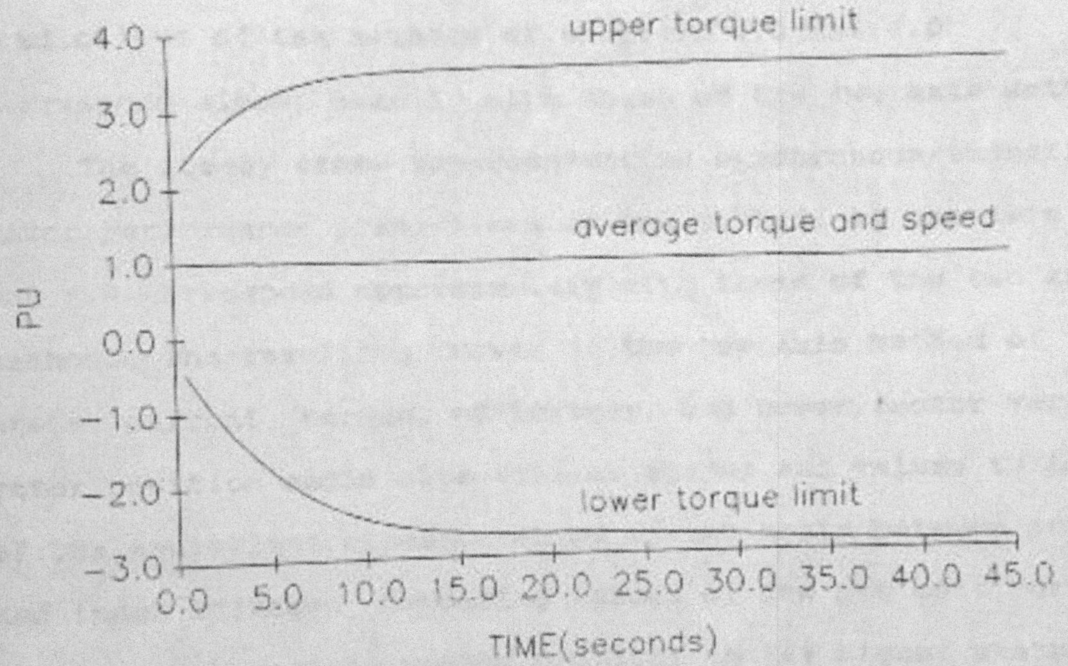

Figure 25 Per unit maximum torque, minimum torque, average torque, and average speed versus time during starting immediately after the transition to superconductivity. Average torque and speed are calculated starting from time = 0 for each average torque curve value. Per unit bases are given in 4.3.2.

7.5 Observations and Conclusions

The two axis theory method makes steady state predictions that are similar to those made with the analysis methods of chapters 4.0, 5.0, and 6.0.

The steady state induction motor performance predictions of the methods of chapters 4.0 and 6.0 correspond almost exactly with those of the two axis method.

The steady state superconducting synchronous/induction motor performance predictions of the methods of chapters 5.0 and 6.0 correspond approximately with those of the two axis method. The resulting curves of the two axis method of stator current, torque, efficiency, and power factor versus rotor position angle have similar shapes and values to those of the equivalent circuit method versus angle between rotor and input voltage. Comparing values at one per unit torque output, the two axis method predicts an 18% higher stator current, a 2% higher efficiency, and a 14% lower power factor relative to the equivalent circuit method and rotating field method values.

The two axis method predicts that speed, torque, and peak current oscillations will increase after transition. The two axis theory doesn't predict that the rotor will come to a steady synchronous speed. It predicts that the rotor speed will increase to a speed that is synchronous only on the average. The rotor will oscillate continuously around synchronous speed. The torque oscillations can be seen in Figure 24 and the peak values of the torque oscillations can be seen in Figure 25.

The speed, torque, and peak current oscillations are similar to what would be seen with an ordinary synchronous machine that has no damping bars. The superconducting rotor with its fixed field flux is a perfectly undamped rotor. It has no provisions for damping like ordinary synchronous machines. It can not lose oscillation energy in the rotor, so it continues to oscillate.

7.6 Program Output Data

7.6.1 Superconducting Synchronous/Induction Motor in the Normal Conduction Mode

```
                                                                    RKS
D-CSMP
INTEGRATION

 TIME          S              EFF            PF             PUTOUT        PUIAS        LAR
     LBR
 0.0000E+00  3.7500E-02  -1.7733E+05   0.0000E+00  -4.8672E-02   0.0000E+00  -6.3501E
-01  -2.0752E-07
 5.0000E-02  3.7500E-02   8.8772E-01   0.0000E+00   7.7425E-01   0.0000E+00  -5.3373E
-01  -3.1683E-01
 1.0000E-01  3.7500E-02   8.9081E-01   0.0000E+00   1.0985E+00   0.0000E+00  -2.3875E
-01  -5.9009E-01
 1.5000E-01  3.7500E-02   8.9067E-01   8.6486E-01   1.1091E+00   1.3366E+00   2.0014E
-01  -6.0549E-01
 2.0000E-01  3.7500E-02   8.8974E-01   8.6441E-01   1.0749E+00   1.3379E+00   5.4232E
-01  -3.2220E-01
```

7.6.2 Superconducting Synchronous/Induction Motor in the Steady State Superconducting Mode

```
                                                                    RKS
D-CSMP
INTEGRATION

 TIME         THRO1         EFF           PF            PUTOUT        PUIAS
 0.0000E+00   2.1000E-01   8.4159E-01   0.0000E+00   1.1563E-01   0.0000E+00
 5.0000E-02   2.1000E-01   9.2273E-01   0.0000E+00   9.1051E-01   0.0000E+00
 1.0000E-01   2.1000E-01   9.2190E-01   0.0000E+00   9.0400E-01   0.0000E+00
 1.5000E-01   2.1000E-01   9.2470E-01   8.1355E-01   9.3453E-01   1.2624E+00
 2.0000E-01   2.1000E-01   9.2267E-01   8.0180E-01   9.0958E-01   1.2650E+00
```

7.6.3 Superconducting Synchronous/Induction Motor in Transition from the Normal Conduction Mode to the Superconducting Mode

D-CSMP RK5
INTEGRATION

TIME	RTIME	PUTOUT	PURPM	PUTAV	RPMAV
0.0000E+00	-1.0000E-01	-1.3084E-01	9.6300E-01	0.0000E+00	0.0000E+00
2.0000E-02	-8.0000E-02	2.6783E-01	9.2606E-01	-3.0239E-01	9.4198E-01
4.0000E-02	-6.0000E-02	1.4333E+00	9.2470E-01	3.2476E-01	9.3265E-01
6.0000E-02	-4.0000E-02	1.5630E+00	9.4000E-01	7.2966E-01	9.3244E-01
8.0000E-02	-2.0000E-02	1.4497E+00	9.5446E-01	9.2472E-01	9.3625E-01
1.0000E-01	-2.2315E-09	1.2232E+00	9.6274E-01	9.9816E-01	9.4081E-01
1.2000E-01	2.0000E-02	2.0110E+00	9.8286E-01	1.1167E+00	9.4580E-01
1.4000E-01	4.0000E-02	2.1693E+00	1.0152E+00	1.2629E+00	9.5341E-01
1.6000E-01	6.0000E-02	1.2265E+00	1.0361E+00	1.3223E+00	9.6267E-01
1.8000E-01	8.0000E-02	1.9036E-01	1.0284E+00	1.2562E+00	9.7071E-01
2.0000E-01	1.0000E-01	-3.1413E-01	9.9507E-01	1.1131E+00	9.7494E-01
2.2000E-01	1.2000E-01	3.8120E-01	9.6562E-01	1.0084E+00	9.7527E-01
2.4000E-01	1.4000E-01	1.6974E+00	9.6651E-01	1.0103E+00	9.7426E-01
2.6000E-01	1.6000E-01	2.1077E+00	9.9425E-01	1.0848E+00	9.7462E-01
2.8000E-01	1.8000E-01	1.9282E+00	1.0252E+00	1.1565E+00	9.7718E-01
3.0000E-01	2.0000E-01	9.0645E-01	1.0379E+00	1.1760E+00	9.8096E-01
3.2000E-01	2.2000E-01	-1.7390E-01	1.0187E+00	1.1228E+00	9.8408E-01
3.4000E-01	2.4000E-01	-1.8476E-01	9.8237E-01	1.0402E+00	9.8506E-01
3.6000E-01	2.6000E-01	7.6631E-01	9.6074E-01	9.9557E-01	9.8417E-01
3.8000E-01	2.8000E-01	1.9906E+00	9.7354E-01	1.0196E+00	9.8314E-01
4.0000E-01	3.0000E-01	2.1952E+00	1.0066E+00	1.0769E+00	9.8345E-01
4.2000E-01	3.2000E-01	1.5797E+00	1.0338E+00	1.1188E+00	9.8527E-01
4.4000E-01	3.4000E-01	5.0920E-01	1.0357E+00	1.1165E+00	9.8764E-01
4.6000E-01	3.6000E-01	-3.6974E-01	1.0066E+00	1.0689E+00	9.8918E-01
4.8000E-01	3.8000E-01	1.0606E-02	9.6990E-01	1.0101E+00	9.8911E-01
5.0000E-01	4.0000E-01	1.3147E+00	9.5998E-01	9.9573E-01	9.8801E-01
5.2000E-01	4.2000E-01	2.1196E+00	9.8376E-01	1.0282E+00	9.8731E-01
5.4000E-01	4.4000E-01	2.2301E+00	1.0186E+00	1.0727E+00	9.8783E-01
5.6000E-01	4.6000E-01	1.1528E+00	1.0400E+00	1.0969E+00	9.8939E-01
5.8000E-01	4.8000E-01	1.0484E-02	1.0294E+00	1.0807E+00	9.9106E-01
6.0000E-01	5.0000E-01	-3.9611E-01	9.9281E-01	1.0350E+00	9.9176E-01
6.2000E-01	5.2000E-01	3.4867E-01	9.6073E-01	9.9742E-01	9.9121E-01
6.4000E-01	5.4000E-01	1.8540E+00	9.6392E-01	1.0010E+00	9.9020E-01
6.6000E-01	5.6000E-01	2.1992E+00	9.9565E-01	1.0349E+00	9.8984E-01
6.8000E-01	5.8000E-01	1.9716E+00	1.0296E+00	1.0691E+00	9.9054E-01
7.0000E-01	6.0000E-01	8.2837E-01	1.0421E+00	1.0797E+00	9.9191E-01

8.0 PROTOTYPE MOTORS

Prototype motors were built to determine first hand some of the problems associated with superconducting high speed induction motors.

No actual superconducting synchronous/induction motors were built.
Motors were built to operate at speeds of up to 60,000 rpm at room temperatures and up to 22,000 rpm at liquid nitrogen temperatures.

Available motor construction materials helped determine which motor designs were tested. For example, the stator tooth and gap geometry were fixed by manufactured stator laminations.

Three stators and twelve rotors were built. The best and most informative of these are reported here.

Figure 26 shows a photograph of a prototype motor in the test setup. Figure 27 shows a photograph of a prototype motor in the test setup without the vacuum flask and instrumentation.

Figure 26 Photograph of a prototype motor in the test setup.

Figure 27 Photograph of a prototype motor in the test setup without the vacuum flask and instrumentation.

8.1 Stator Design

Three stators were built. One of these was used only in preliminary testing and is not reported here. The other two stators were built from identical materials to the same dimensions, but with different windings. A photograph of the discussed stators is shown in Figure 28. A drawing of the stators without their windings is shown in Figure 29.

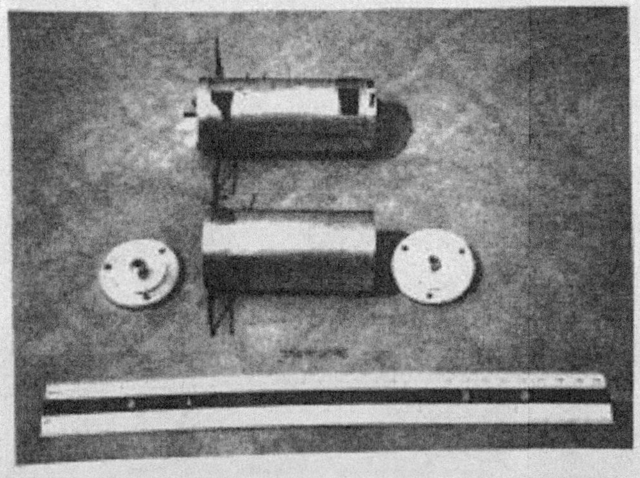

Figure 28 Photograph of the stators. The upper has a 4 pole two phase winding. The lower has an 8 pole two phase winding.

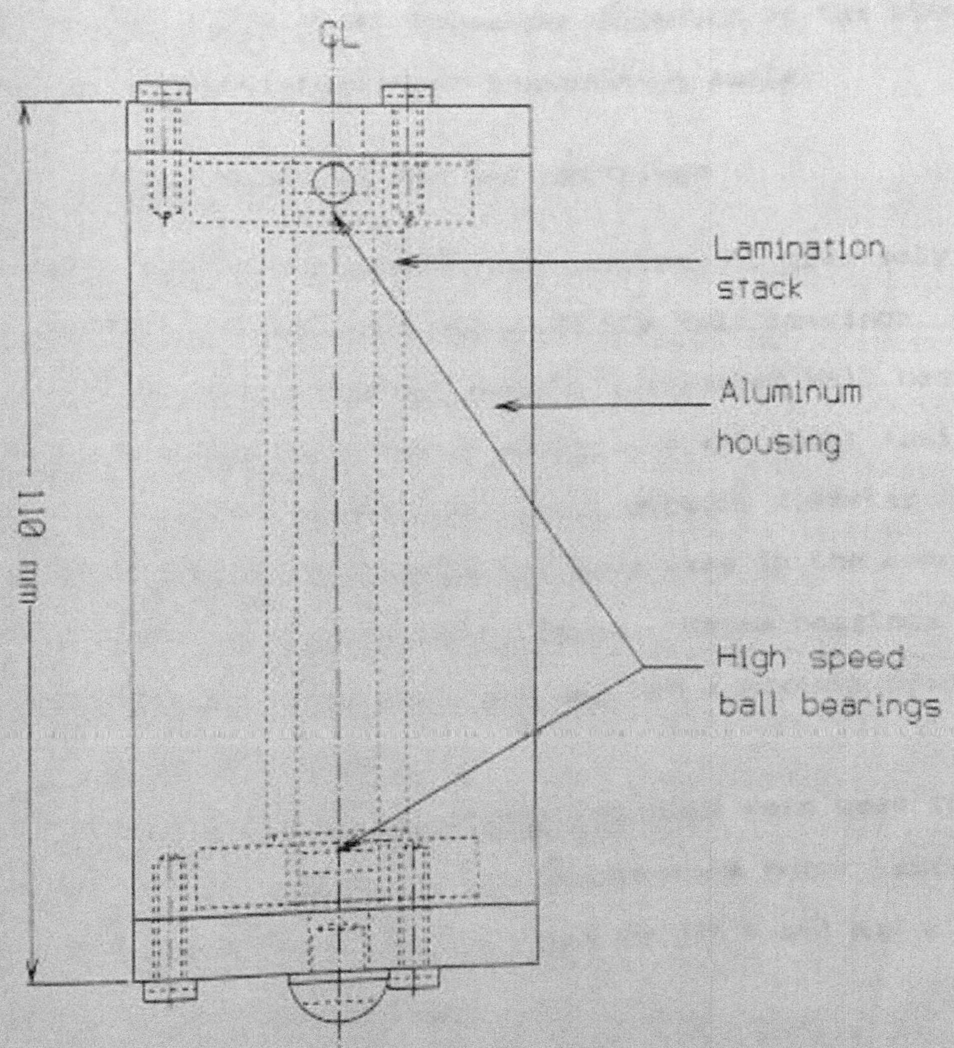

Figure 29 Prototype stator. Windings are not shown.

8.1.1 Stator Housings

The stator housings were turned on a precision lathe from stock cylinders of aluminum. Aluminum was chosen for its easy machinability, nonferromagnetic nature, and heat conducting ability. The outside diameter of the housing was kept at the larger than necessary diameter of the stock aluminum to make handling and machining easier.

8.1.2 Stator End Bells and Ball Bearings

End bells were turned from aluminum to precisely fit into the stator housings and hold the ball bearings.

Grease lubricated and teflon lubricated ball bearings were used. They each had a .00318 m (.125 inch) inside diameter and a .00635 m (.25 inch) outside diameter. The grease lubricated ball bearings were used in the motors during room temperature motor tests. Those bearings were rated for room temperature use and had a maximum speed rating of 60,000 rpm.

The teflon lubricated ball bearings were used in the motors during liquid nitrogen temperature motor tests. Those bearings were rated for use at 77° K and had a maximum speed rating of 22,000 rpm.

8.1.3 Stator Laminations

Stator laminations having sixteen slots were donated by the Magnetic Metals Corporation. The laminations are Magnetic Metals type 596ST. They are made of Superperm 49. Their thickness is .000356 m (.014 inches). A drawing of a lamination is shown in Figure 30.

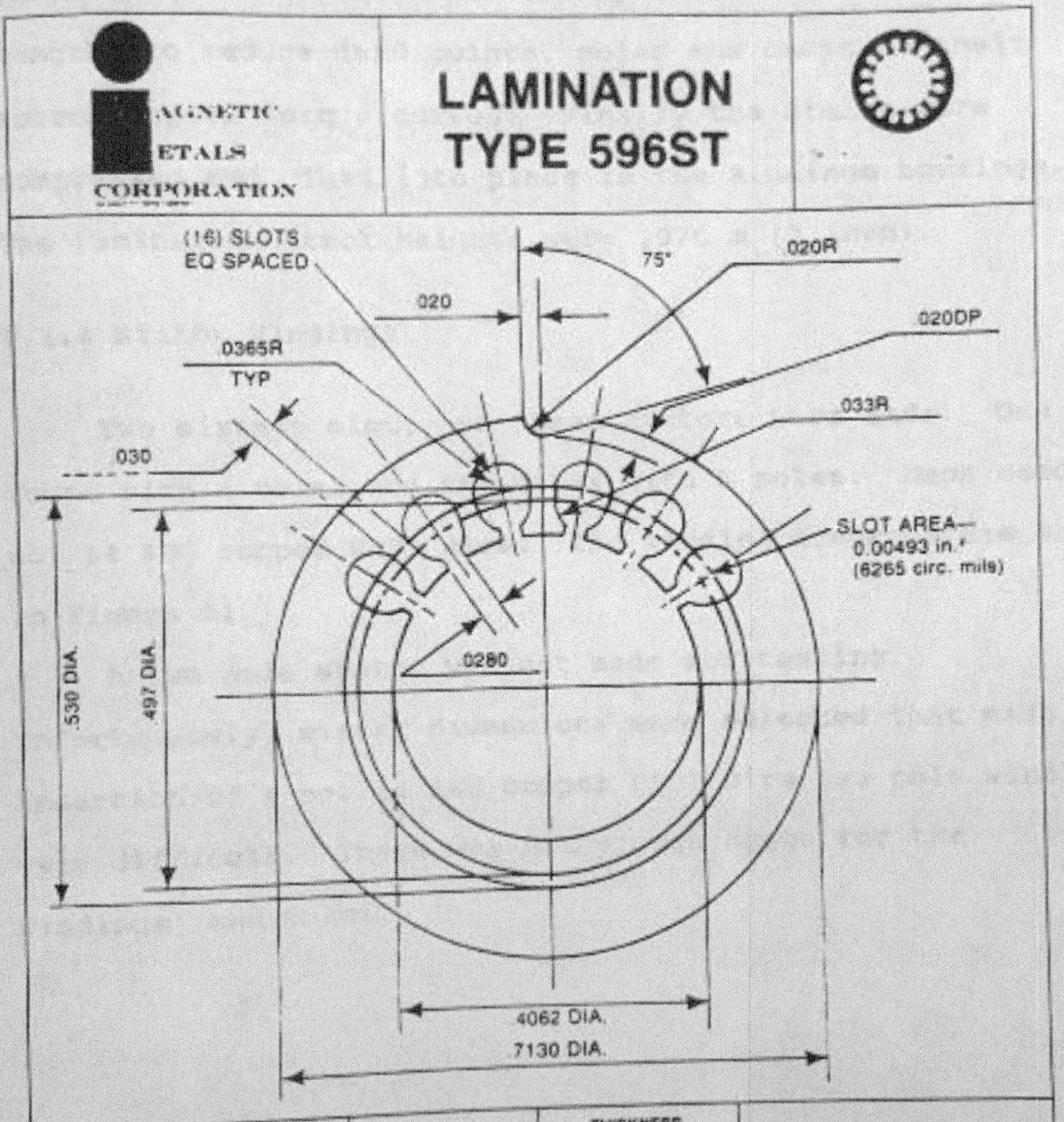

MAGNETIC METALS MATERIAL DESIGNATION	MATERIAL CODE	THICKNESS (inches)	CATALOG NUMBER
SUPERPERM 49	43	.010	596ST4310
SUPERPERM 49	43	.014	596ST4314
SUPERPERM 80	80	.010	596ST8010
SUPERPERM 80	80	.014	596ST8014
NOMEX	NX	.010	596STNX10

Figure 30 Stator lamination. Drawing courtesy of the Magnetic Metals Corporation.

Stator laminations were made into stacks. Then the laminations were skewed one slot pitch over their stack lengths to reduce dead points, noise and cusps in their motors' speed torque curves. Finally the stacks were compressed and glued into place in the aluminum housings. The lamination stack heights were .076 m (3 inch).

8.1.4 Stator Windings

Two sixteen slot, two phase stators were made. One was wound with 4 poles and the other with 8 poles. Each used no. 14 AWG copper HAPT wire. The winding diagrams are shown in Figure 31.

A two pole stator was not made for testing. Unfortunately, stator dimensions were selected that made the insertion of a no. 14 AWG copper HAPT wire two pole winding very difficult. There was not enough space for the windings' end turns.

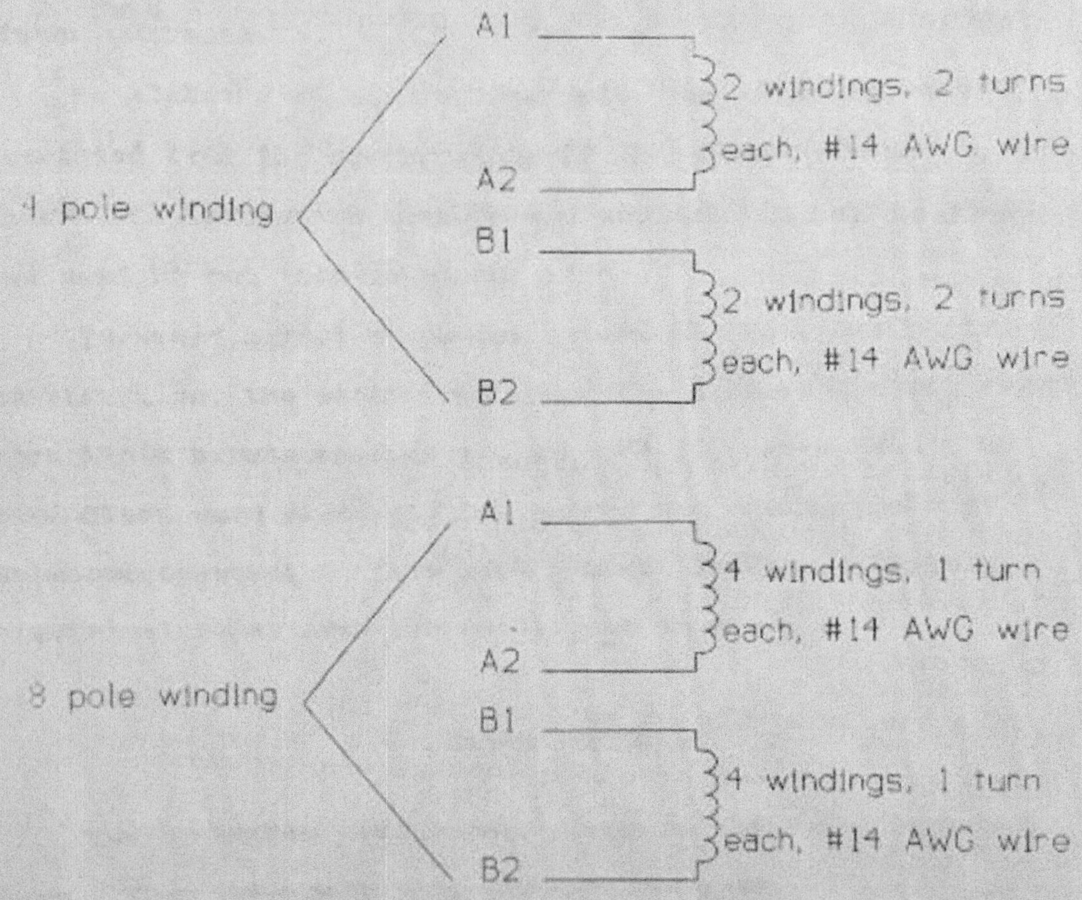

Figure 31 Winding diagrams of the four and eight pole stator windings.

The stators were made with a single conductor per slot. This was done here, as it was done in the work of Hesmondhalgh and Tippings[15], to maximize the amount of conductor material in each slot. Maximizing conductor material minimized resistive losses in the stator. With only a single conductor per slot higher currents were required. Step down transformers were used to supply the higher currents.

No stator slot liners were used, the windings were insulated from the stator slots by the wires' own poly-thermalene insulating coating and Stycast blue epoxy glue was used to pot them in place.

To avoid scraping insulation off of the wires during construction, the stator was wound one slot at a time. Each slot had a single conductor slid into it. After all conductors were slid into the slots, the conductors were soldered together to form each phases' winding. Ordinary electrical 60/40 lead tin solder was used.

8.2 Rotor Designs

Twelve rotors were built. Four of these are reported here. They represent four distinct designs.

The rotors did not have identical outside diameters. The radial rotor to stator air gap varied from .00013 m (.0053 inches) for the steel jacketed squirrel cage rotor to .00031 m (.012 inches) for the ordinary squirrel cage rotor. A larger air gap requires a larger magnetizing current. Larger magnetizing currents cause more stator copper losses. By subtracting the stator copper losses away from the input powers comparisons can be made between the different rotor types.

The rotors are shown in a photograph in Figure 32.

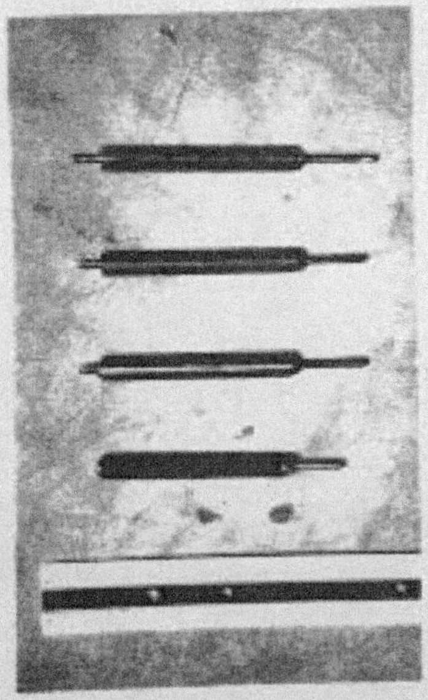

Figure 32 Photograph of the rotors. They are from top to bottom the solid rotor, the laminated squirrel cage rotor, the steel jacketed laminated squirrel cage rotor, and the laminated rotor without a squirrel cage.

8.2.1 Solid Rotor

The solid rotor is the simplest to construct and the most rugged. Solid rotors have been made and evaluated by many induction motor researchers. Occasionally they have been used in industry as high starting torque, short operating time induction motors. Some of the research done on high speed, solid rotor induction motors was reported by Lakerveld[13], Petit[18], and Ikeda[19].

The solid rotor and was turned on a precision lathe from a solid bar of low carbon Ryerston 1018 steel.

A drawing of the solid rotor is given in Figure 33.

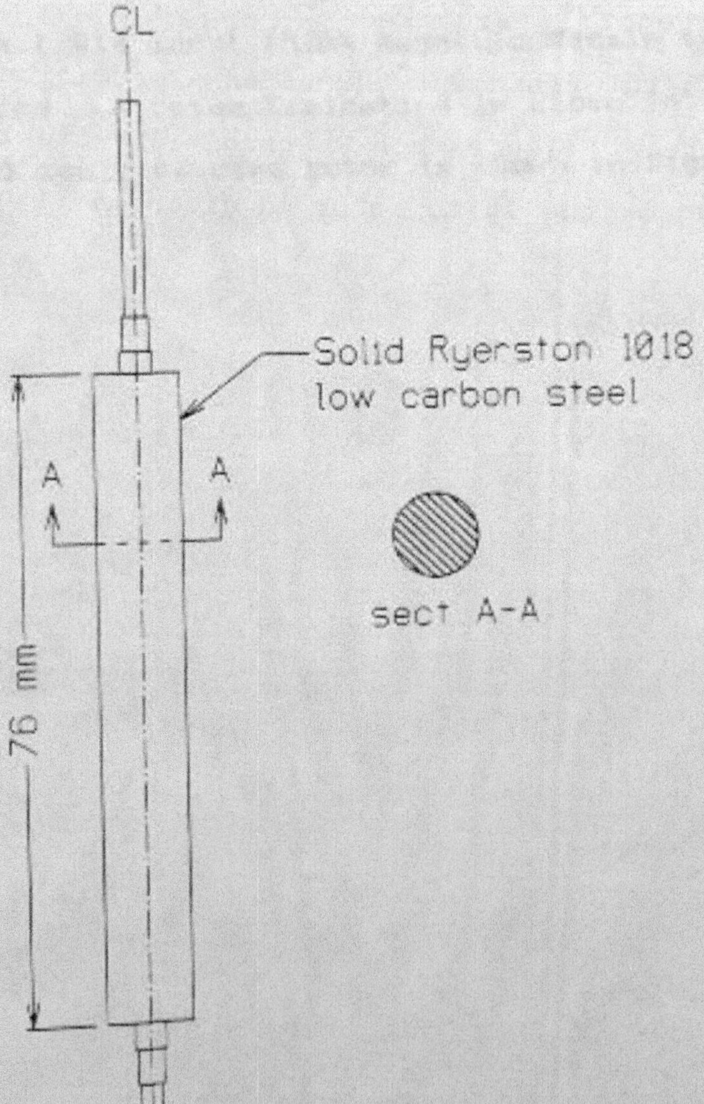

Figure 33 Solid rotor.

8.2.2 Laminated Squirrel Cage Rotor

A squirrel cage rotor was made of copper bars and a stack of rotor laminations. The rotor laminations were donated by the Magnetic Metals Corporation. They are .000356 m (.014 inch) thick Magnetic Metals type 183RR rotor laminations. A rotor lamination is shown in Figure 34. The laminated squirrel cage rotor is shown in Figure 35.

MAGNETIC METALS CORPORATION

LAMINATION TYPE 183RR

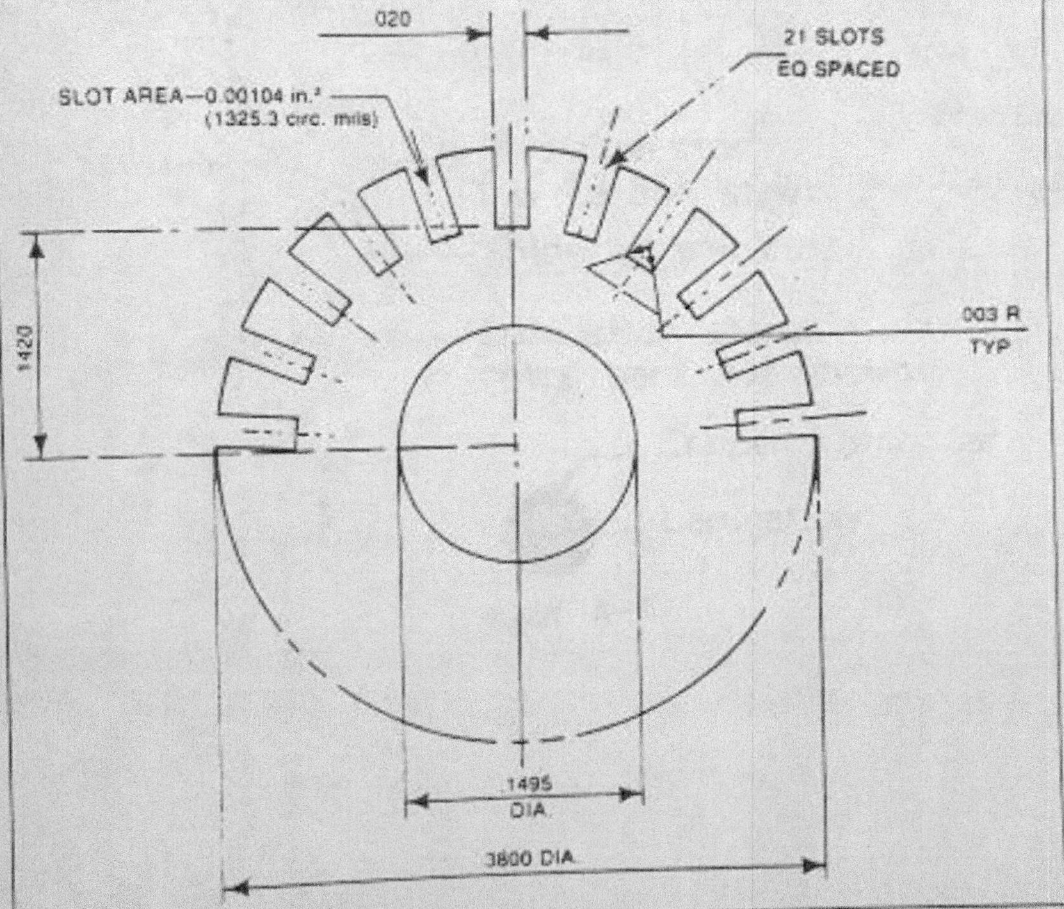

MAGNETIC METALS MATERIAL DESIGNATION	MATERIAL CODE	THICKNESS (inches)	CATALOG NUMBER
SUPERPERM 49	43	.014	183RR4314

Figure 34 Rotor lamination. Drawing courtesy of the Magnetic Metals Corporation.

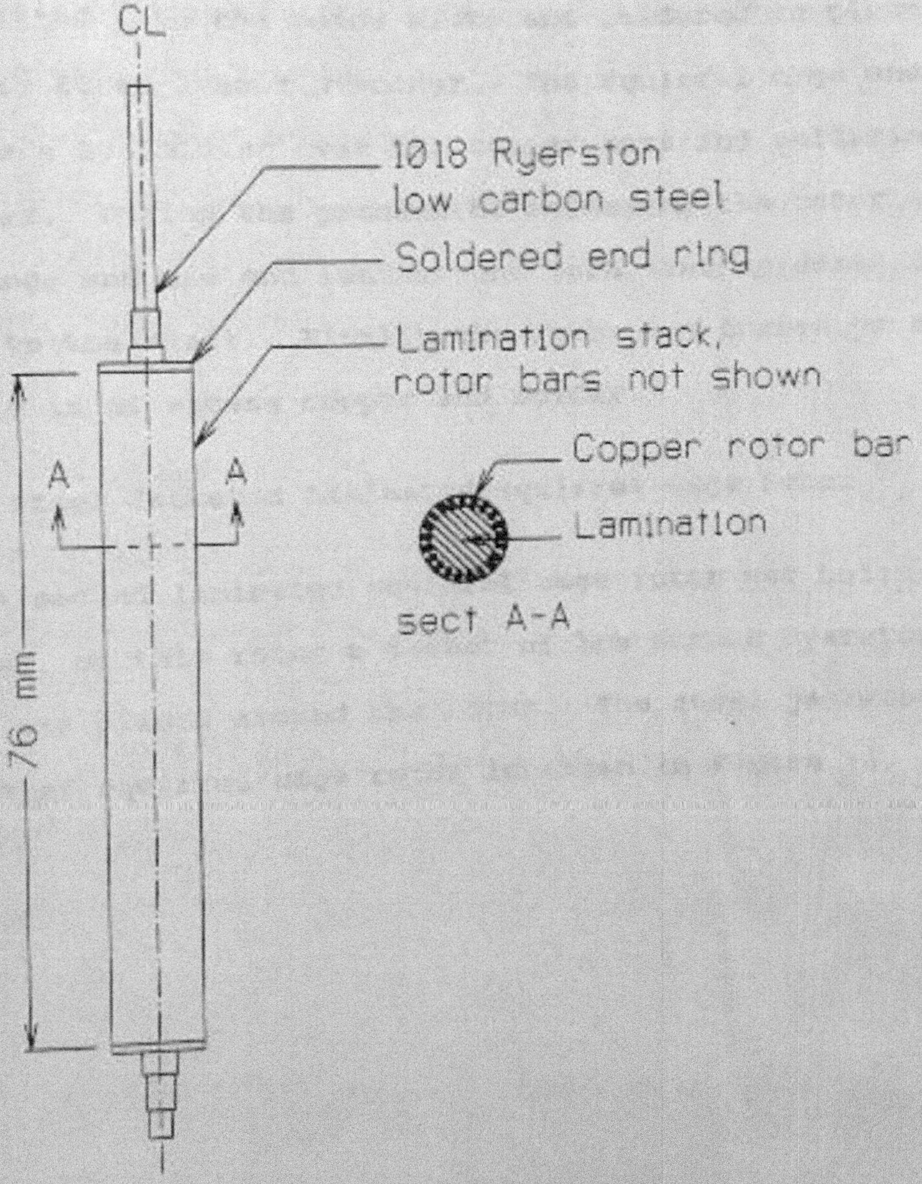

Figure 35 Laminated squirrel cage rotor.

The first step in motor construction was the creation of the shaft. Low carbon Ryerston 1018 steel was turned to shape on a precision lathe. Then a stack of rotor laminations was slid over the shaft. Next the rotor bars were fitted into the rotor slots and soldered in place with ordinary 60/40 lead tin solder. The squirrel cage end rings were made by folding over the copper bars and soldering them together. During the process of soldering the rotor bar end-rings and the end laminations were also soldered in place to the shaft. Finally the rotor was turned on a lathe to trim it of excess copper and solder.

8.2.3 Steel Jacketed Laminated Squirrel Cage Rotor

A second laminated squirrel cage rotor was built. However, on this rotor a jacket of low carbon Ryerston 1018 steel was placed around the rotor. The steel jacketed laminated squirrel cage rotor is shown in Figure 36.

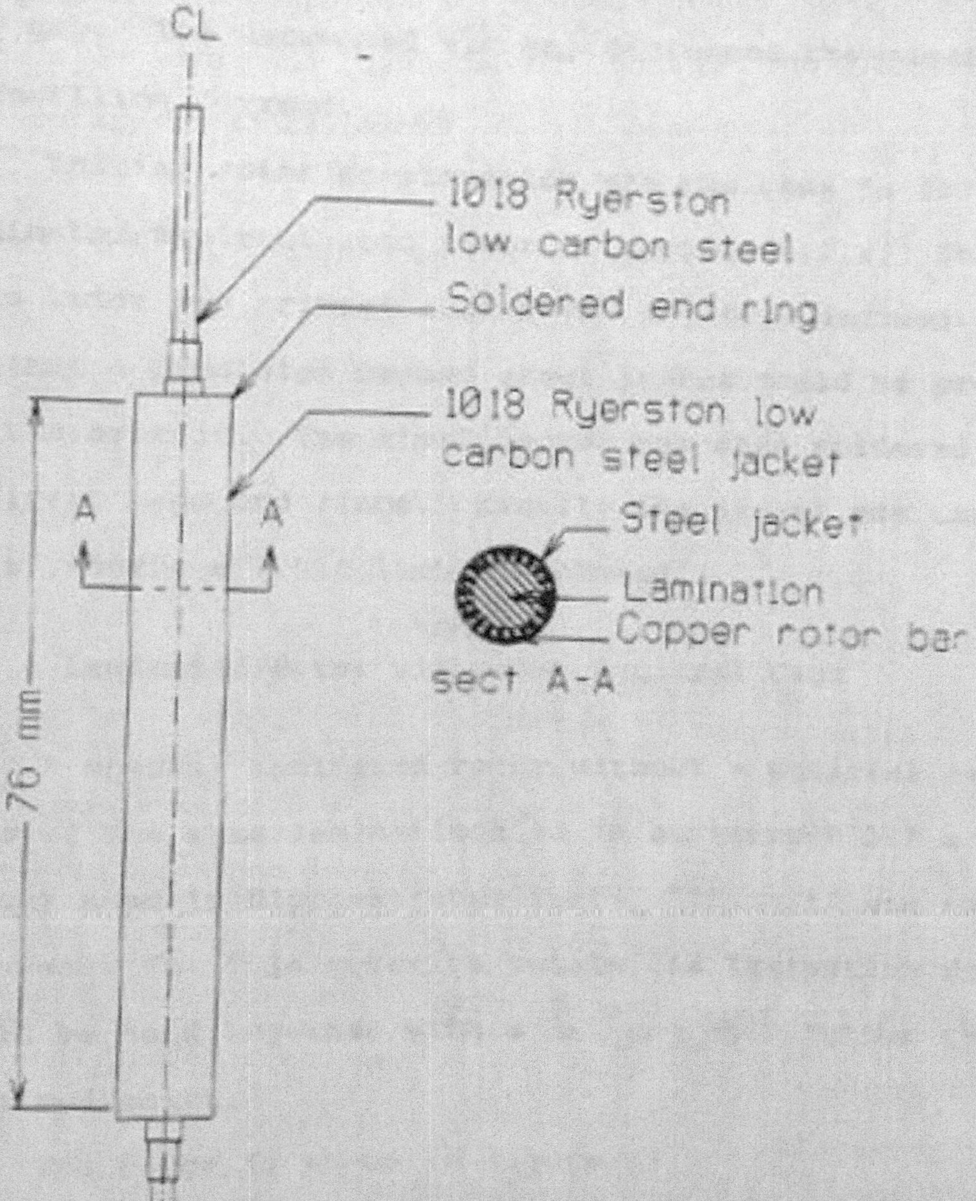

Figure 36 Steel jacketed laminated squirrel cage rotor.

The steel jacket performed two functions. First, it guaranteed that the rotor bars could not be forced out of the rectangular rotor slots by centrifugal force during high speed operation. Second, it decreased the rotor to stator air gap. The decreased air gap decreased the required magnetizing current.

Initial rotor construction was the same as for the laminated squirrel cage rotor of section 8.2.2. Then the bare rotor was trimmed 6.35×10^{-5} m (.0025 inches) radially so that a precision reamed steel jacket could be press fitted over it. The steel jacket was then soldered to the squirrel cage end rings. Finally the jacket was turned down to a .000254 m (.010 inch) thickness.

8.2.4 Laminated Rotor without a Squirrel Cage

A special laminated rotor without a squirrel cage was made of the same laminations as in sections 8.2.2 and 8.2.3. It was used in blocked rotor tests. Since it was not necessary for this rotor to rotate its lamination stack could be held together with a nut and bolt rather than a machined shaft.

The rotor is shown in Figure 37.

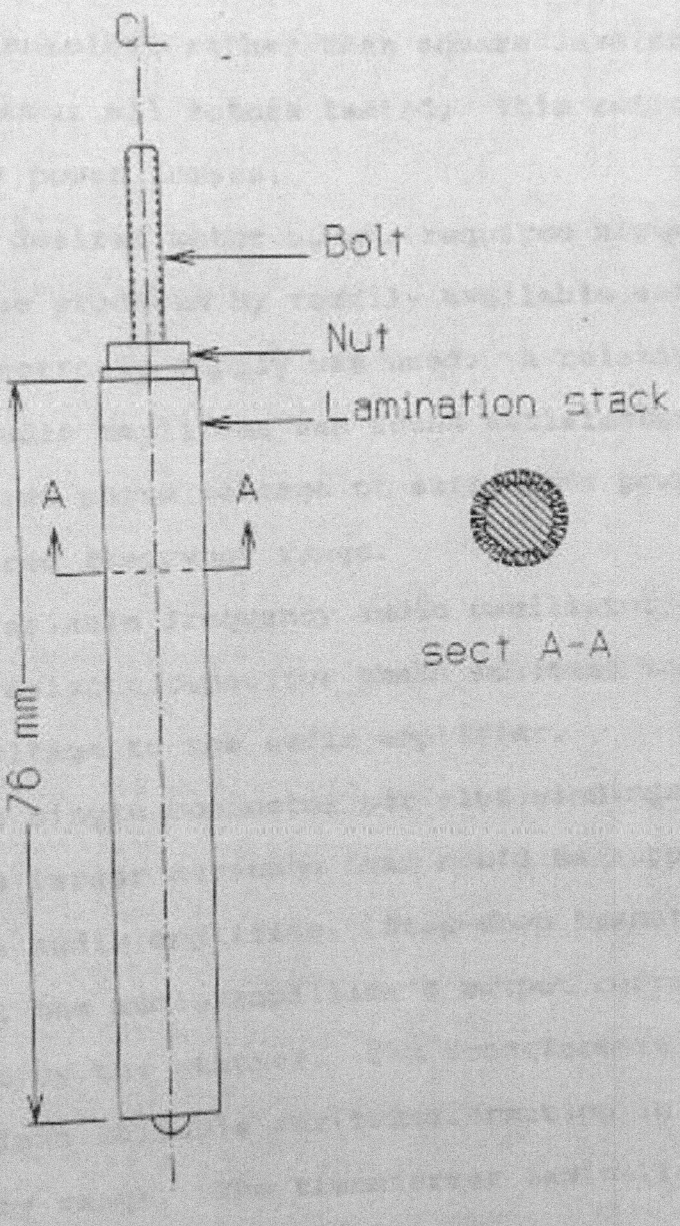

Figure 37 Laminated rotor without a squirrel cage.

8.3 Auxiliary Equipment

8.3.1 Power Supply

In accords with Lakerveld's[13] paper it was decided that a sinusoidal rather than square wave source would be used to power all motors tested. This reduced harmonic frequency power losses.

The desired motor speeds required higher frequencies than those produced by readily available alternators, so a power electronic supply was used. A relatively high power stereo audio amplifier was found satisfactory. It could produce two phase voltage of sufficient power and quality in the desired frequency range.

A variable frequency audio oscillator output went to a simple resistor/capacitor phase splitter to supply two phase input voltage to the audio amplifier.

The single conductor per slot windings of the stators required larger currents than could be supplied directly from the audio amplifier. Step-down transformers were used to boost the audio amplifier's output current to the levels required by the stators. The transformers used thin laminations suitable for transformation in the desired frequency range. The transformer laminations were donated by the Magnetic Metals Corporation.

A schematic of the power supply circuit and instrumentation is shown in Figure 38.

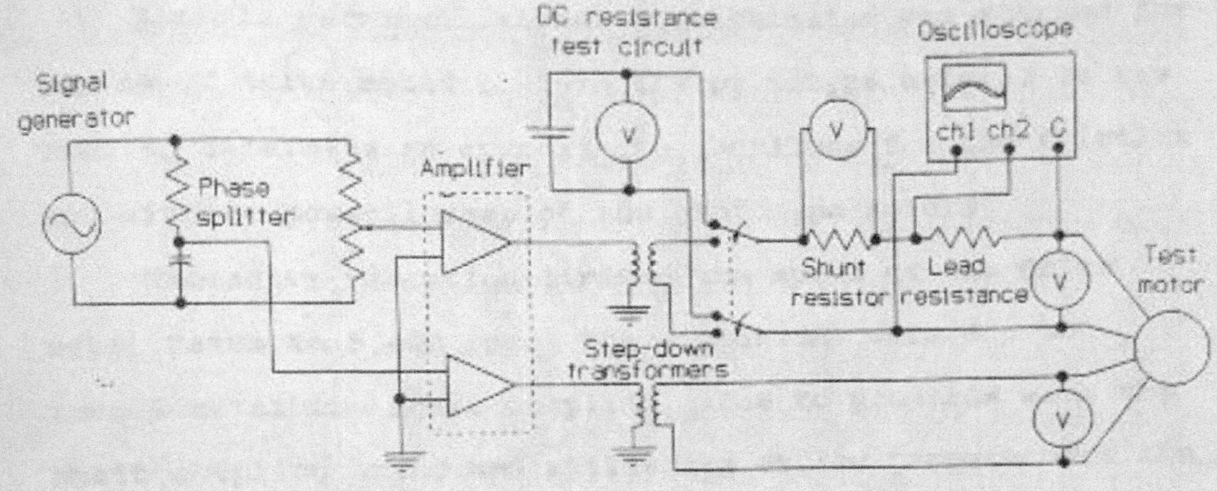

Figure 38 Schematic of the motor power supply circuit and instrumentation.

8.3.2 Cooling System

For the low temperature tests, cooling of the prototype motors was achieved by simply immersing them in liquid nitrogen. The vertical stator was immersed up to the level of its output shaft end bell in liquid nitrogen. The level of the liquid nitrogen was kept below the end bell's output shaft opening so that no liquid nitrogen actually flowed into the motor. Liquid nitrogen was held in a vacuum flask. The vacuum flask can be seen in the photograph of Figure 26.

For the room temperature tests the motors were cooled with the surrounding air.

8.3.3 DC Drive Motor

A small permanent magnet DC tachometer was adapted for use as DC drive motor to turn the prototype motors. It was used to determine an approximate magnitude for the friction and windage power losses of the prototype motors.

Excessive vibration limited the speed of the drive motor tests to 5,400 rpm. The vibrations were due to imbalance at the shaft coupling. Due to problems with the shaft coupling icing and stiffening at low temperatures the drive motor could only be used during room temperature testing.

The DC drive motor can be seen mounted above a prototype motor in the photograph of Figure 39.

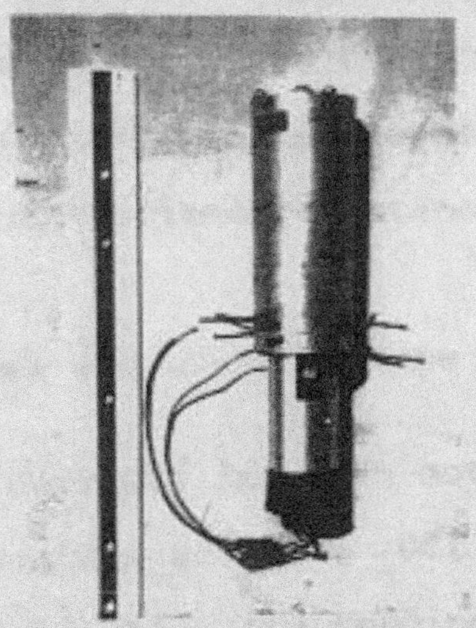

Figure 39 Photograph showing the DC drive motor mounted on a prototype motor.

8.3.4 Warming Fan

During the liquid nitrogen tests between torque readings a small fan was used to blow room temperature air across the motor shaft and eddy current disk. This helped reduce icing on those parts. It can be seen in Figure 26.

8.3.5 Instrumentation

8.3.5.1 <u>Voltage</u>. Phase voltage into the motor was measured directly with DVM's.

Only one phase's voltage was used. However, the power supply circuit could accidentally produce different voltage magnitudes on each phase. Therefore, the other phase's voltage was also checked and adjusted as necessary.

Since the motor used low voltage and high currents, care had to be taken in locating the connections for voltage measurement. The voltmeters were connected at the motor's input leads to eliminate lead resistance voltage drop inaccuracies.

Figure 38 shows where the voltage readings were taken.

8.3.5.2 Current. Currents into the motors were determined by measuring the voltage across a .001 Ω meter shunt with a DVM. Figure 38 shows where the readings were taken.

8.3.5.3 Current Lag. The lag of the phase current relative to phase voltage into the motor was determined with a dual trace oscilloscope. The current waveform was measured with one oscilloscope trace as the voltage across a short length of one phase's input lead near the motor input. The voltage was measured across the same phase with the other oscilloscope trace.

Figure 38 shows where the current lag readings were taken.

8.3.5.4 Stator Winding Resistance and Temperature. Stator DC winding resistance was determined with the voltmeter/ammeter method applied to one phase. A high current double throw double pole knife switch allowed quick switching of one phase from the ordinary AC phase supply to a DC supply. Resistance is DC volts over DC amps.

Knowledge of the liquid nitrogen temperature resistance of the copper winding and the equation for copper resistivity change as a function of temperature allowed the temperature of the stator winding to be calculated in the liquid nitrogen tests.

Figure 38 shows the resistance measuring circuit for DC stator winding resistance testing.

8.3.5.5 Speed. Speed was measured with a hand held strobe. It can be seen in Figure 26.

8.3.5.6 Torque During Rotation. A device using the principles of the eddy current brake and the beam balance was made to measure the motor output torque. A drawing of this device is shown in Figure 40. The device is also shown in the photographs in Figures 26 and 27.

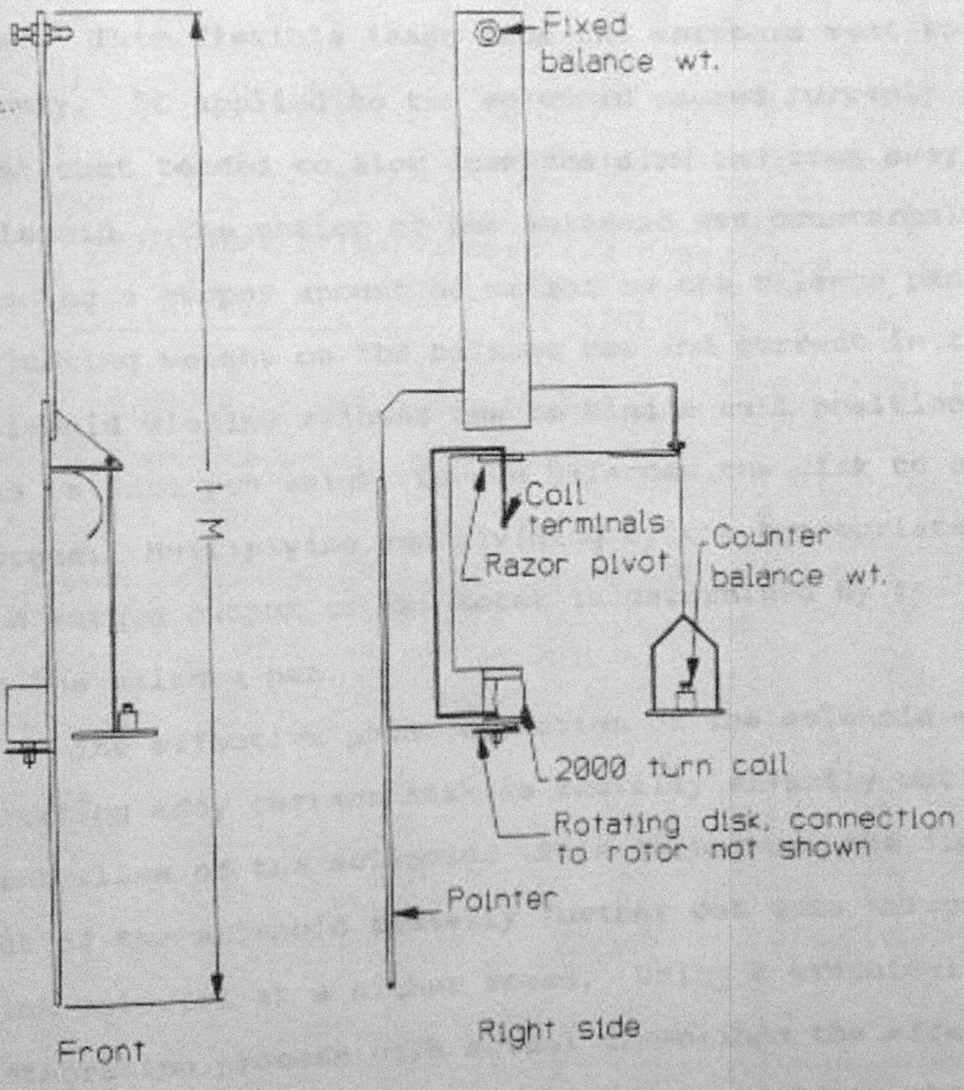

Figure 40 Rotating rotor torque measuring device.

There were several parts to the device. An aluminum eddy current disk was attached to the shaft of the motor. Above the disk a 2000 turn iron core solenoid was hung. The solenoid was hung on a structure that could pivot on a razor edge, like a beam balance. It allowed the solenoid free movement over the disk in a circumferential direction on the disk. Thin flexible leads from the solenoid went to a DC supply. DC applied to the solenoid caused currents in the disk that tended to slow down the disk and push away the solenoid. The motion of the solenoid was counterbalanced by placing a proper amount of weight on the balance pan. Adjusting weight on the balance pan and current in the solenoid winding allowed one to find a null position where the balance pan weight torque balances the disk to solenoid torque. Multiplying and dividing by the appropriate radii the torque output of the motor is determined by the weight on the balance pan.

The effective point of action of the solenoid on the rotating eddy current disk is radially slightly out from the centerline of the solenoid. This is because the flux going out of the solenoid radially further out goes through the disk material at a higher speed. Using a graphical integration process with actual dimensions the effective point of action was determined to be 1.06 times the radial distance between the center of the rotating disk and the centerline of the solenoid.

The rotating eddy current disk was heat insulated from the motor shaft by a single layer of plastic electrical tape. This layer was necessary to reduce icing during the liquid nitrogen temperature tests.

The rotating eddy current disk with its relatively large radius and layer of plastic electrical tape shaft insulation was the greatest source of imbalance in the motor load tests. Tests at higher speeds than 30,000 rpm should use a smaller diameter disk and be attached to the shaft over a more precisely fitted shaft insulator.

8.3.5.7 Blocked Rotor Shaft Clamp. Blocked rotor torque was measured with a clamp on the rotor's shaft and a spring balance. Force perpendicular to the clamp was measured with a hand held spring balance. Force times radius gave the torque. The device is shown in Figure 41.

Figure 41 Photograph showing the blocked rotor torque measuring clamp and spring balance.

8.3.5.8 <u>Accuracy</u>. There were many potential sources of inaccuracy. These varied from less than 3% to up to 30% absolute accuracy. The most accurate were the phase voltage measurements, the least were the torque measurements. To confound the matter, the degree of accuracy varied with magnitude of the quantity being measured. For example, high torques could be much more accurately determined than low torques.

Data obtained during testing was generally used in a relative way. For example, the magnetic losses were determined versus frequency over a range of frequencies, with the same instruments being used in the same way for each series of readings. Accuracy of one reading relative to another is better than absolute accuracy.

An in-depth error analysis was not done. It is felt it would add little to the research. However, care was taken by the author to not draw conclusions beyond the accuracy limitations of the instrumentation.

8.4 Experimental Tests and Graphed Results

Tests were done to explore the effects of frequency, rotor design, stator design, and temperature on induction motor performance.

The prototype motors have larger than usual stator copper power losses. As was explained by Miller[23] this is because their rotor to stator air gaps are large relative to their winding cross-sections. They require high magnetizing currents and have relatively high resistance stator windings. To make the data more useful for comparison with more usual motors and with each other the stator copper losses have been subtracted from the total input powers in many of the following graphs.

8.4.1 Room Temperature Tests

8.4.1.1 Stator Resistance and Resistance Power Losses. Stator resistance was determined with the DC voltmeter/ammeter method. A DC current was applied to one phase of the motor and the voltage drop measured across the stator winding. Stator resistance is by Ohms law. No attempt was made to compensate for the skin effect, the frequencies used in the motor were too small to warrant its consideration.

Stator resistance power losses were calculated as:

$$P_{Rs} = 2I^2R_s \tag{8-1}$$

While operating at room temperature the stator winding resistance per phase was measured as .009 Ω per phase for the 4 pole stator and .0087 Ω per phase for the 8 pole stator.

8.4.1.2 No Load, Rotor Running Free Tests. No load tests were done versus supply frequency for a constant motor input of 17 VA. The constant input VA indicates what the motor can do while receiving a maximum input. In each test currents, voltages, current lag angles, and motor speeds were recorded. Other testing procedures, such as those that keep the stator voltage constant, internal stator voltage constant or the stator voltage to supply frequency ratio constant, were not used.

Tests were done for each of the three rotors in each of the two stators.

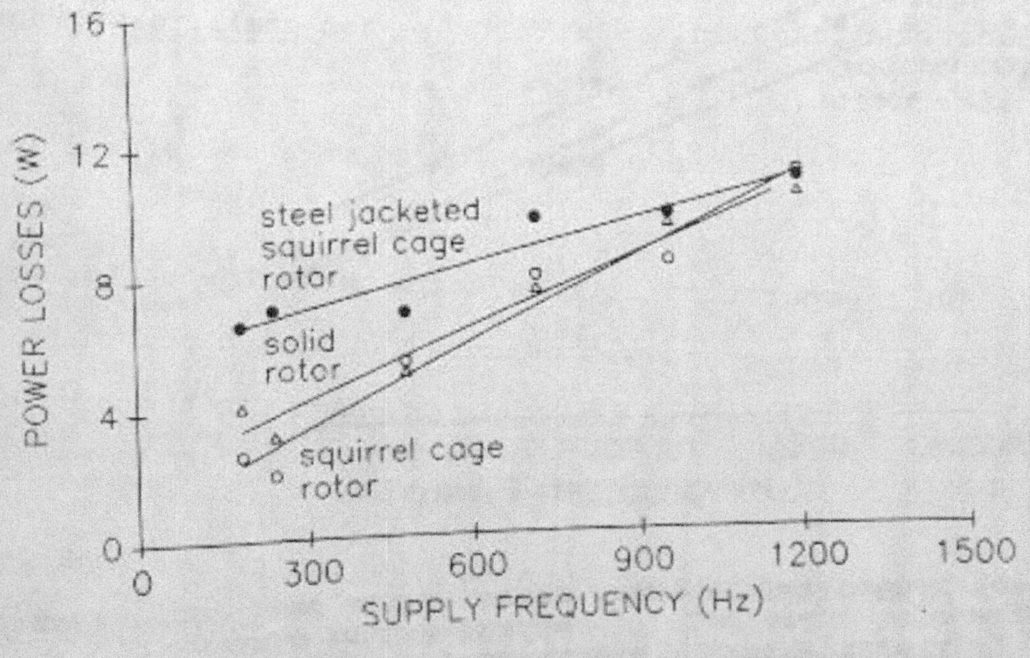

Figure 42 No load power losses minus stator copper losses versus supply frequency. The four pole motor was used at room temperature. Approximately 17 VA were input into the motors for each data point. The lines are first order least square approximations.

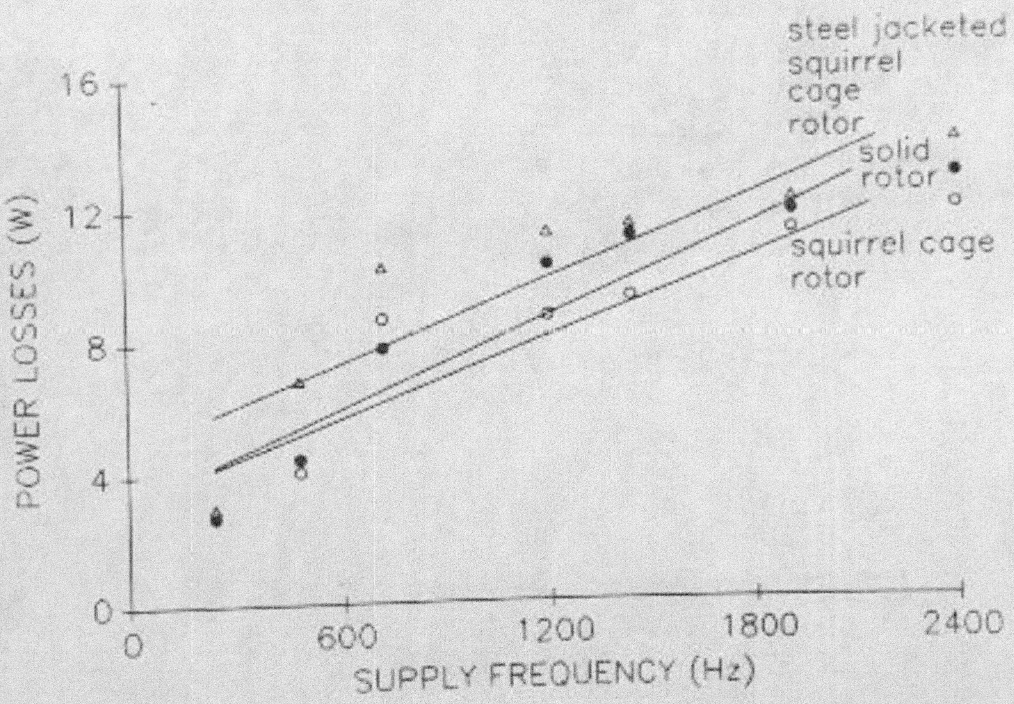

Figure 43 No load power losses minus stator copper losses versus supply frequency. The eight pole motor was used at room temperature. Approximately 17 VA were input into the motors for each data point. The lines are first order least square approximations.

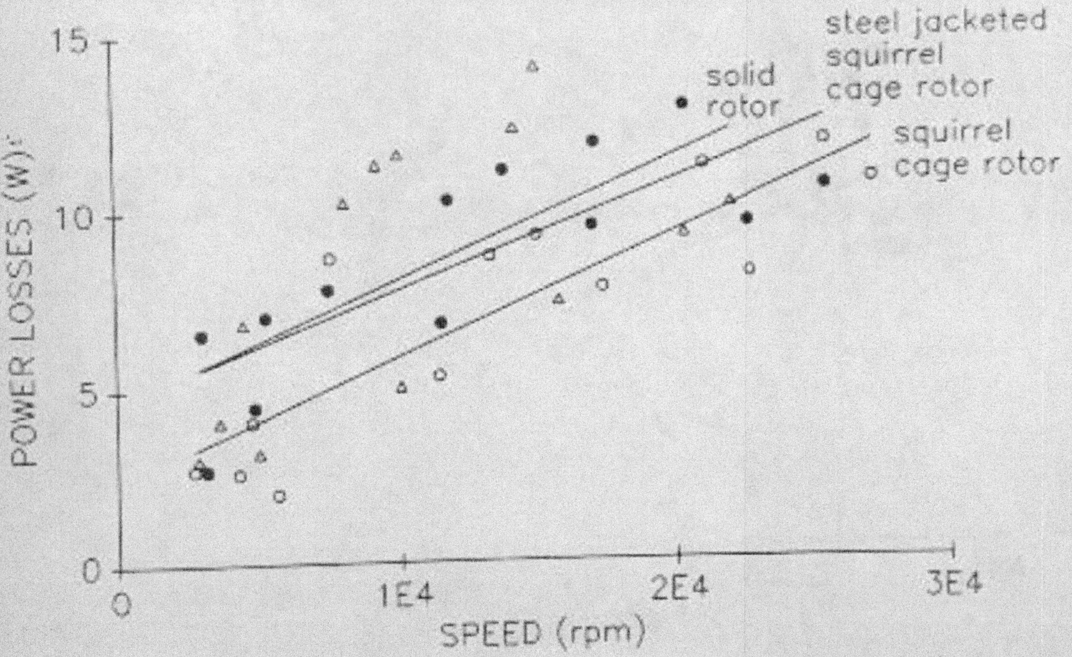

Figure 44 No load power losses minus stator copper losses versus speed as the supply frequency was varied. Data from the room temperature 4 pole and 8 pole motors were combined for each rotor type. Approximately 17 VA were input into the motors for each data point. The lines are first order least square approximations.

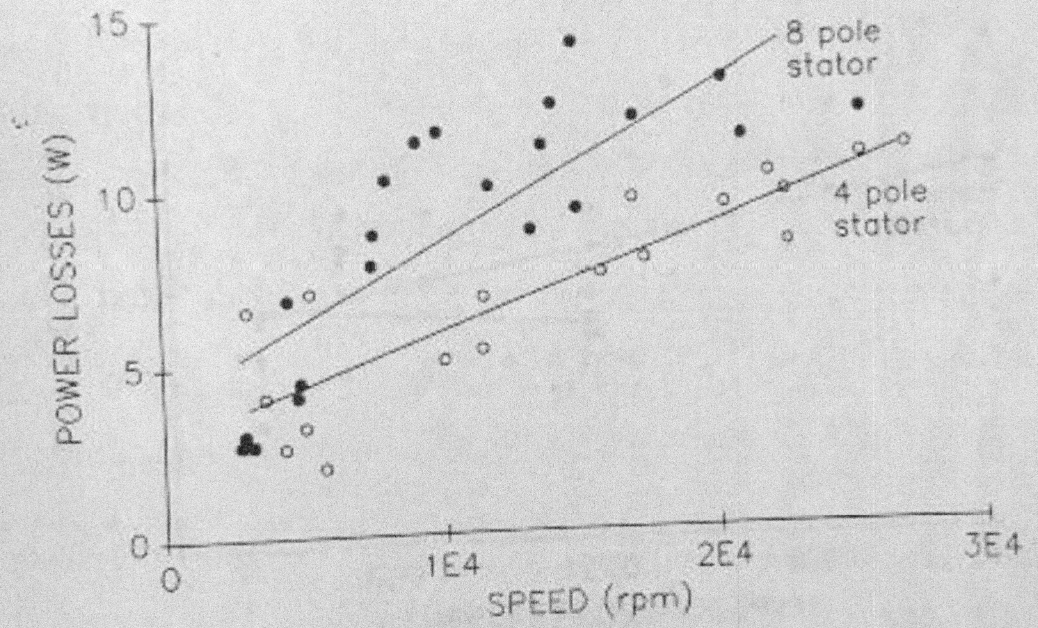

Figure 45 No load power losses minus stator copper losses versus speed as the supply frequency was varied. Data from each rotor type were combined into groups for room temperature 4 pole and 8 pole motors. Approximately 17 VA were input into the motors for each data point. The lines are first order least square approximations.

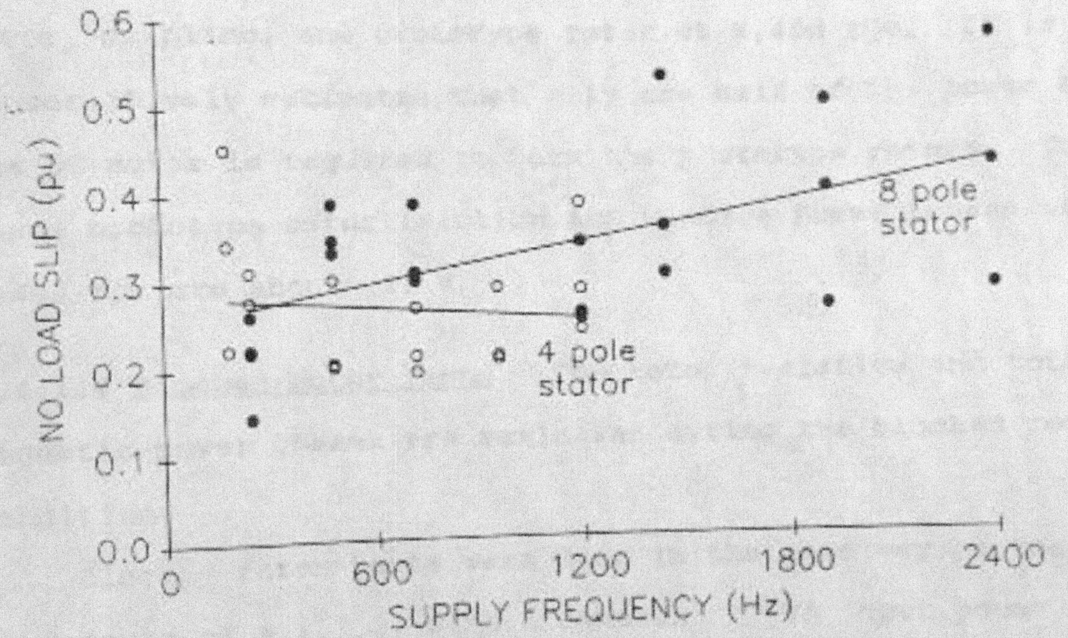

Figure 46 No load slip as the supply frequency was varied. Data from each rotor type was combined into groups for room temperature 4 pole and 8 pole motors. Approximately 17 VA were input into the motors for each data point. The lines are first order least square approximations.

8.4.1.3 DC Drive Motor. A DC motor was used to drive the different rotors at 5,400 rpm to determine friction and windage power losses. Due to excessive vibrations, caused by the shaft coupling, tests at higher speeds could not be done.

The DC motor used an average of .58 W to drive the DC motor, coupling, and prototype rotor at 5,400 rpm. It is conservatively estimated that only one half of the power to the DC motor is required to turn the prototype rotors. This means prototype motor friction and windage power losses at 5,400 rpm are about .29 W.

8.4.1.4 Blocked Rotor Tests. The rotor resistive and total magnetic power losses are maximized during the blocked rotor condition.

Blocked rotor tests were done in the same way as the no load tests of 8.4.1.2, with a constant 17 VA input power for a range of frequencies. From this the total magnetic power losses and the resistive power losses were obtained. As in 8.4.1.2, the input powers were subtracted by the stator resistive power losses.

Blocked rotor tests were done on the squirrel cage rotor and the rotor composed of a stack of laminations without a squirrel cage.

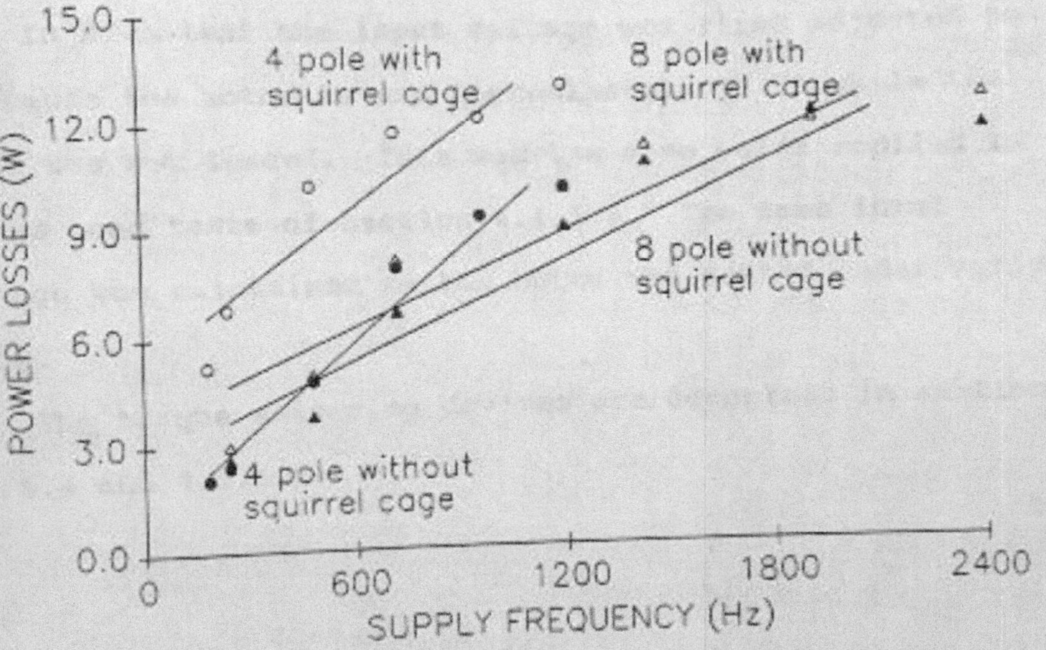

Figure 47 Blocked rotor power losses minus stator copper losses as supply frequency was varied. Approximately 17 VA were input into the room temperature motors for each data point. Motors used either the squirrel cage rotor or the laminated rotor without a squirrel cage as indicated. The lines are first order least square approximations.

8.4.1.5 Load Tests. Load tests were done for a constant motor input voltage applied to the 4 pole stator with the different rotors. In each test currents, voltages, current lag angles, motor speeds, and torques were recorded.

The squirrel cage rotor tests were done at 240, 720, and 1,200 Hz. The other two rotors were tested only at 720 Hz. In each test the input voltage was first adjusted to the cause the motor to use approximately 17 VA while the motor was not loaded. This was the same value applied in the no load tests of section 8.4.1.2. The same input voltage was maintained as the motor was tested under various loads.

The torque measuring devices are described in sections 8.3.5.6 and 8.3.5.7.

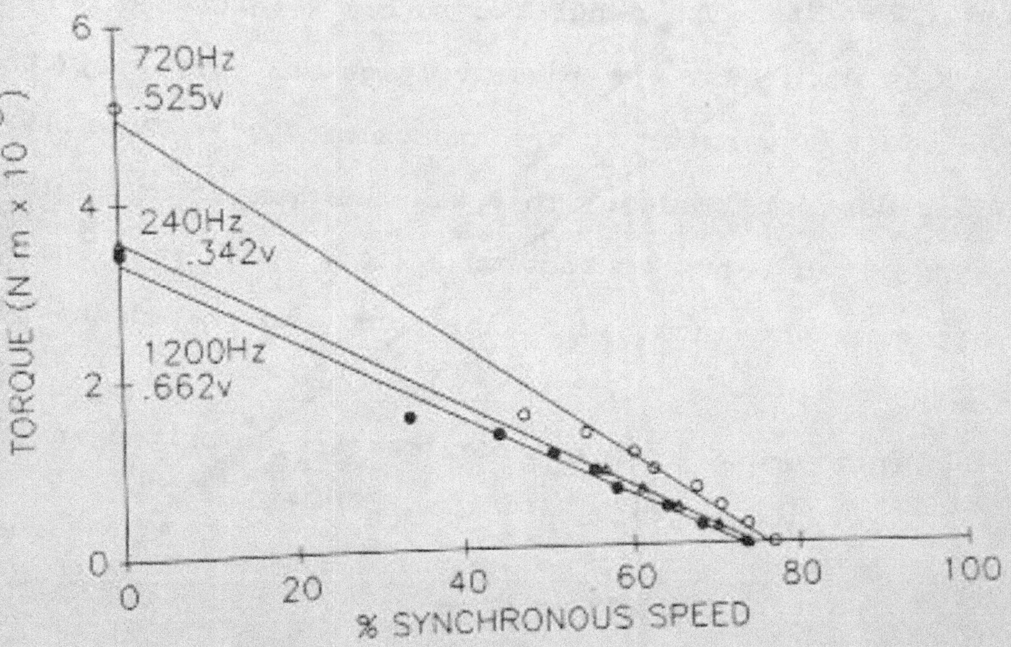

Figure 48 Output torque versus per cent synchronous speed for the room temperature 4 pole stator with the squirrel cage rotor. The lines are first order least square approximations.

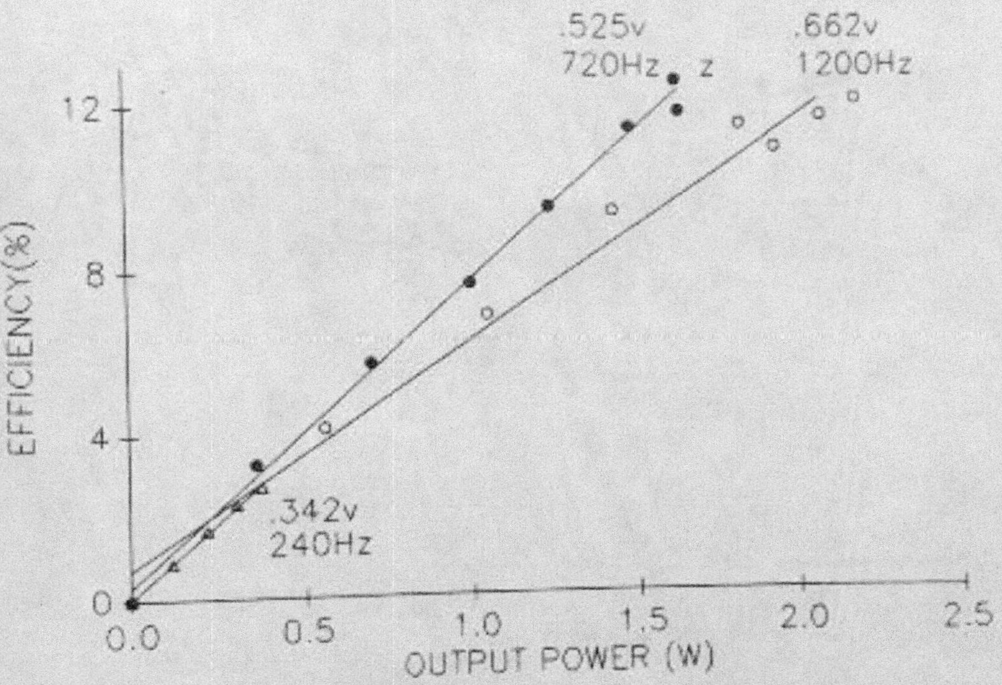

Figure 49 Output per cent efficiency versus output power for the room temperature 4 pole stator with the squirrel cage rotor. The lines are first order least square approximations.

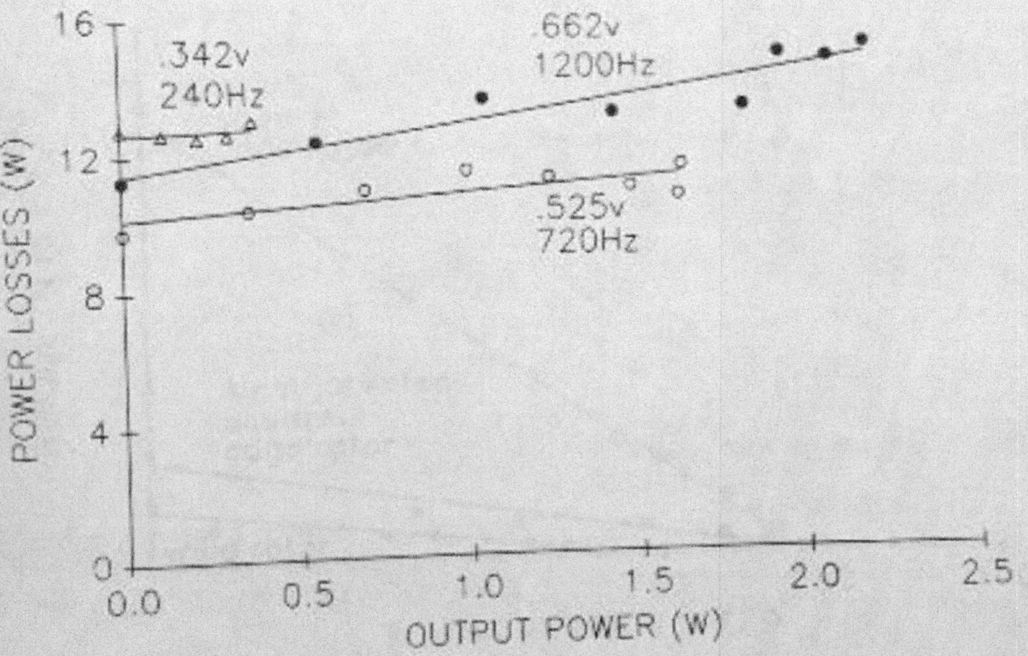

Figure 50 Power losses versus output power for the room temperature 4 pole stator with the squirrel cage rotor. The lines are first order least square approximations.

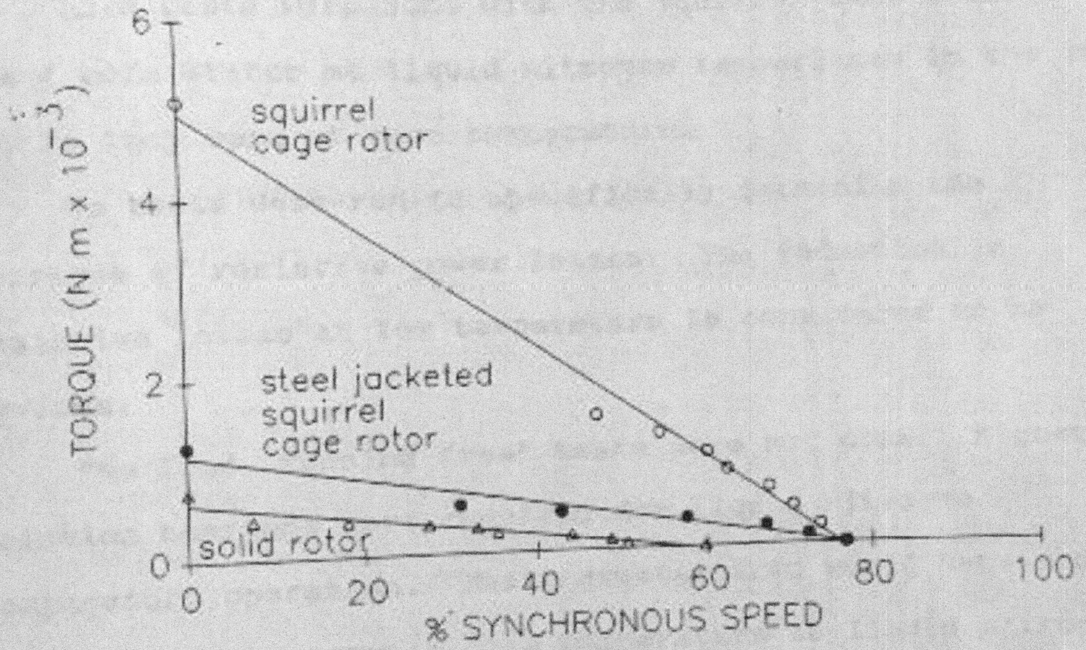

Figure 51 Output torque versus per cent synchronous speed for the room temperature 4 pole motors. 720 Hz were input into each motor for each point. The lines are first order least square approximations.

8.4.2 Liquid Nitrogen Temperature Tests

Magnetic losses at liquid nitrogen temperatures were determined with blocked rotor tests. Using the 4 pole stator, tests were done on the squirrel cage rotor and the laminated rotor without a squirrel cage.

Load tests were done with the squirrel cage rotor in the 4 pole stator at liquid nitrogen temperature in the same way as they were at room temperature.

No tests were run to specifically determine the decrease of resistive power losses. The reduction in resistive losses at low temperature is considered to be obvious.

"No load, running free" tests were not done. Higher friction bearings were required for liquid nitrogen temperature operation. Their greater drag would have made it difficult to compare room temperature to liquid nitrogen temperature performance.

8.4.2.1 Stator Resistance, Resistance Power Losses, and Temperature. Stator resistance and resistive power losses were determined here as they were in 8.4.1.1.

The DC resistance values were also used with the resistance as a function of temperature equation of copper to calculate the stator winding's temperature. The equation used to determine temperature is taken from the Standard Handbook for Electrical Engineers[45]. It is:

$$\frac{R_{t2}}{R_{t1}} = \frac{234.4 + t_2}{234.4 + t_1} \tag{8-2}$$

Liquid nitrogen temperature stator resistance varied between .00115 Ω at liquid nitrogen temperature, 77° K, (-196° C) to .00157 Ω after a 16 W power loss for 10 minutes while the stator is immersed in liquid nitrogen. The resistance during the 16 W power loss indicates the stator winding's temperature is 91° K (-182° C) in that condition.

8.4.2.2 Blocked Rotor Tests. Blocked rotor tests were done with the 4 pole stator on the squirrel cage rotor and the laminated rotor without a squirrel cage. The same stator input currents and frequencies were used as in the room temperature blocked rotor tests of 8.4.1.4. The use of the same currents made the comparison of room temperature to liquid nitrogen temperature test results easier.

8.4.2.3 Load Tests. Load tests were done for the squirrel cage rotor in the 4 pole stator. In each test a constant motor input voltage was applied. Current, voltage, current lag angle, motor speed, and torque were recorded.

The squirrel cage rotor tests were done with 240, 720, and 1,200 Hz. In each test the input voltage was first adjusted to the cause the motor to use approximately 17 VA while the motor was not loaded. The same input voltage was maintained as the motor was tested under load.

The torque measuring devices are described in sections 8.3.5.6 and 8.3.5.7.

Icing was a significant problem. All cooled metal parts exposed to the room air condensed and froze water from the air. Even with the warming air fan, see section 8.3.4, the motors were only good for satisfactory operation for about 25 minutes before they seized up. It is believed that the seizing was caused by the output shaft bearing icing up.

8.4.3 Graphical Comparison of Liquid Nitrogen Temperature Tests to Room Temperature Tests

8.4.3.1 Blocked Rotor Effective Motor Inductive Impedance versus Frequency.

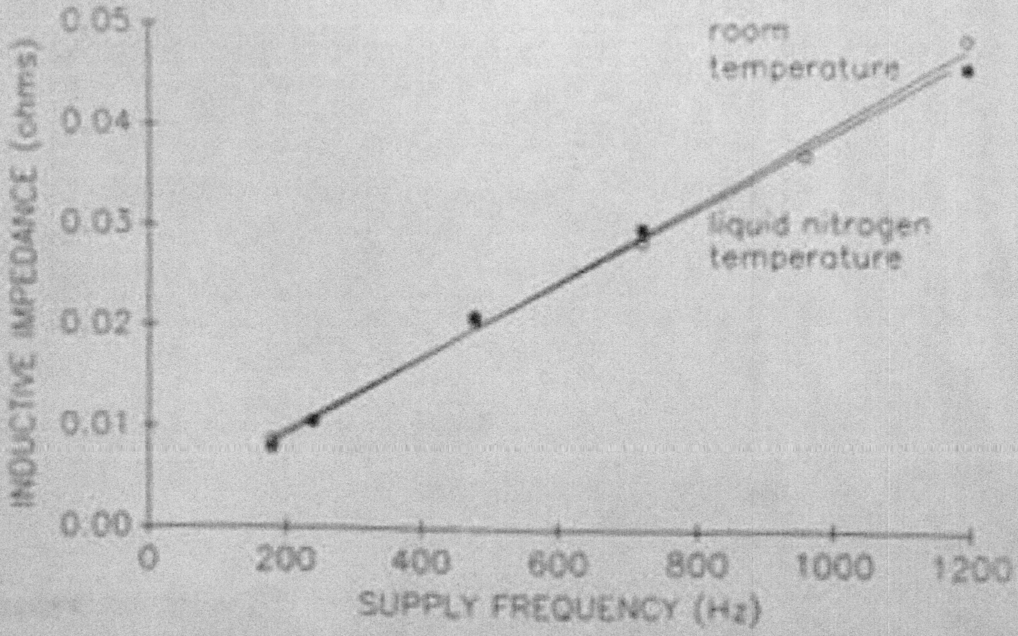

Figure 52 Blocked rotor effective input inductive impedance as supply frequency was varied for the laminated rotor without a squirrel cage. The same input current values were used for each data point at each frequency. The lines are first order least square approximations.

8.4.3.2 Blocked Rotor Power Losses versus Frequency.

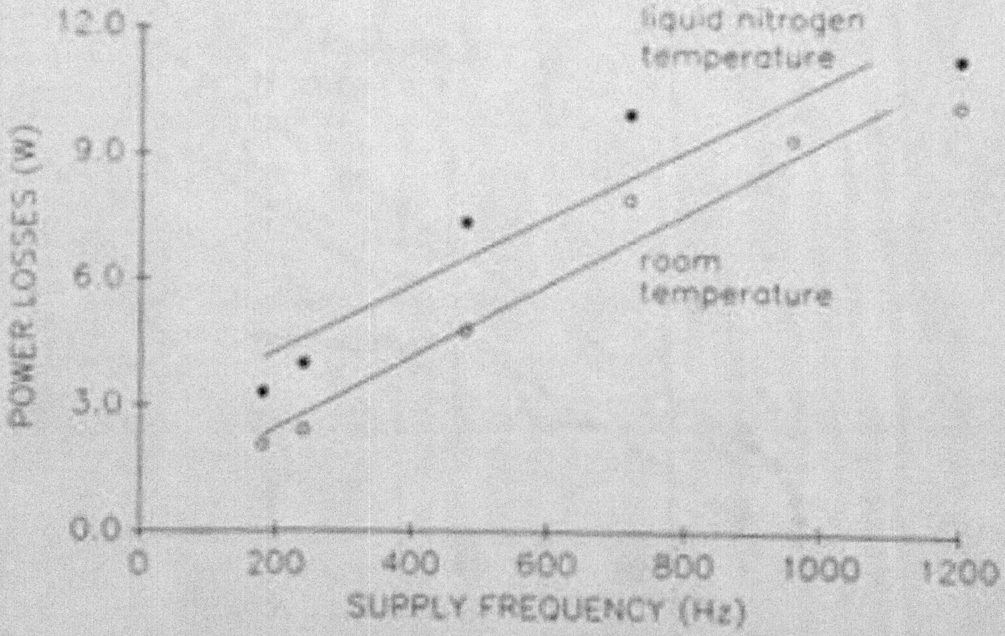

Figure 53 Blocked rotor power losses minus stator copper losses as supply frequency was varied for the laminated rotor without a squirrel cage. The same input current values were used for each data point at each frequency. The lines are first order least square approximations.

8.4.3.3 <u>Output Torque versus % Synchronous Speed</u>.

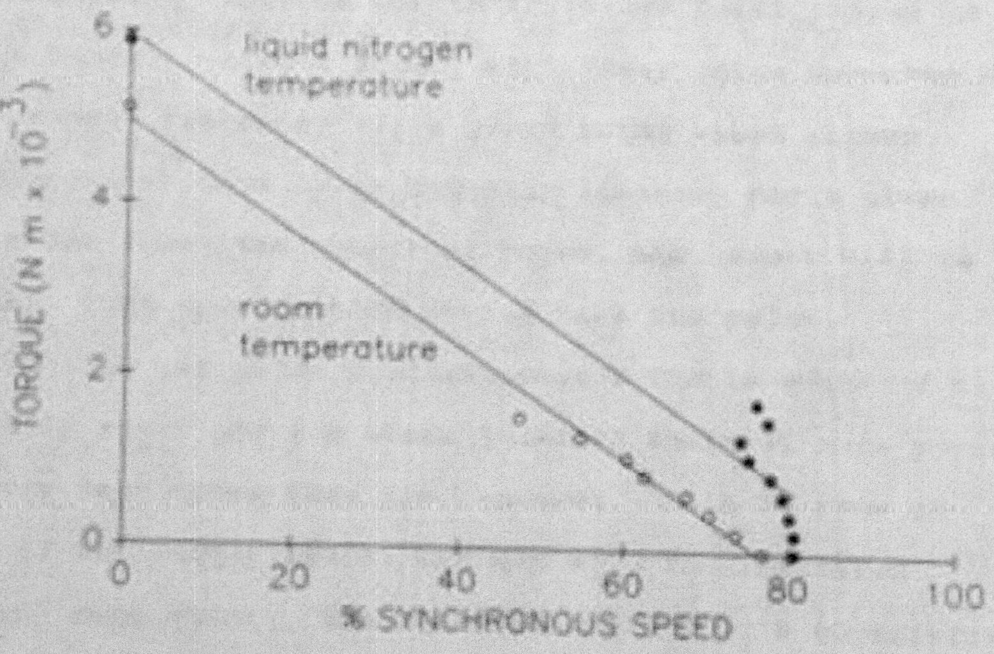

Figure 54 Output torque versus per cent synchronous speed for the 4 pole stator with the squirrel cage rotor. 720 Hz were input into each motor for each point. The lines are first order least square approximations.

8.5 Observations and Conclusions

Many observations can be made about and conclusions drawn from the construction and testing. The following were selected to be most useful to one desiring to improve on the motor designs presented.

1) The 4 pole stator is superior to the 8 pole stator in every test. This is due to increased tooth losses in the 8 pole stator. Also a stator with fewer poles requires a lower supply frequency for a given rotor speed, lower frequencies produce lower magnetic losses. For a given stator the fewer the number of poles, the lesser will be the losses. High speed motors should have two poles.

2) The laminated squirrel cage rotor is superior to the solid rotor and the steel jacketed squirrel cage rotor in every test where they are compared. This is true in spite of the larger radial air gap with the laminated squirrel cage rotor. The squirrel cage rotor's superiority is primarily due to reduced eddy current losses in the squirrel cage laminated rotor. Secondary reasons for this are the increased permeability and decreased hysteresis losses of the rotor laminations compared to low carbon steel. Laminated squirrel cage rotors should be used on high speed induction motors as they are on normal speed induction motors.

3) The stator copper losses were substantial. They

were the major cause of inefficiency in the motors, see section 1.4.5. To improve this the rotor to stator air gap should be minimized. Losses could also be decreased by using silver wire and using wire that better conformed to the stator slots. Also, when using lower supply frequencies, the slot size might be increased to allow more conductor material.

4) There appears to be an optimum supply frequency of about 720 Hz for the 4 pole stator and squirrel cage rotor configuration. This is the result of an ideal balance between the motor's stator resistive losses and magnetic losses. The stator resistive losses decrease with frequency while the magnetic losses increase with frequency.

5) The inductive impedance of the motor composed of the 4 pole stator and the unlaminated rotor does not change as the motor is cooled from room temperature to liquid nitrogen temperature. This is seen in the plot of inductive impedances in the room temperature and liquid nitrogen temperature blocked rotor tests in Figure 52. This indicates that the magnetic properties of the laminations do not change significantly as they are cooled to liquid nitrogen temperatures.

6) When the prototype motor is immersed in liquid nitrogen coolant and is receiving 16 W of waste heat its stator windings appear to experience only a 14° C

temperature rise, according to the DC voltmeter/ammeter test. However, in the liquid nitrogen temperature blocked rotor test a power loss was experienced that was greater than that predicted assuming only a 14° C stator winding temperature rise, see Figure 53. It is believed that the stator windings while carrying AC current are more than 14° C warmer than the liquid nitrogen and that they rapidly cool to the temperature of their surrounding laminations as soon as the stator winding is disconnected from its AC input. It is the surrounding laminations that are 14° C warmer than liquid nitrogen during motor operation. The rotor experiences the same 14° C temperature rise as the laminations. For the present high T_c superconductors this 14° C rise is low enough to allow superconductivity. Therefore the prototype motor design would be suitable for testing high T_c superconductor rotors, when materials with high enough current, magnetic, and frequency capacity become available.

7) Icing of bearings and motor parts is a significant problem. The prototype motor and its eddy current disk could only be used for about 25 minutes before seizing up. If longer usage of the motor is desired a method permitting less icing should be devised.

8) Overhung rotor laminations should be used. The laminations used in the squirrel cage rotors had rectangular

slots. Although no trouble was experienced with the rectangular slots, the centrifugal forces on the bars were substantial. Overhung rotor bar slots would help eliminate the catastrophic failure that would occur if a rotor bar worked loose.

8.6 Tabulated Test Data

8.6.1 Room Temperature Tests

8.6.1.1 No Load, Rotor Running Free Tests.

4 pole stator
squirrel cage rotor
approximately 17 VA constant input

f(Hz)	rpm	input voltage (V)	input current (A)	phase lag (deg)	output torque (N m)
180.	4,200.	.337	26.	32.4	0.0
240.	5,620.	.342	25.	38.9	0.0
480.	11,450.	.444	20.	44.9	0.0
720.	17,400.	.525	16.	41.5	0.0
960.	22,800.	.602	14.	44.9	0.0
1,200.	27,300.	.662	13.	34.6	0.0

4 pole stator
steel jacketed squirrel cage rotor
approximately 17 VA constant input

f(Hz)	rpm	input voltage (V)	input current (A)	phase lag (deg)	output torque (N m)
180.	2,950.	.443	20.	38.9	0.0
240.	5,200.	.495	17.	43.2	0.0
480.	11,500.	.671	13.	55.3	0.0
720.	17,000.	.845	10.	46.7	0.0
960.	22,700.	.845	10.	46.7	0.0
1,200.	25,600.	1.046	8.	43.2	0.0

4 pole stator
solid rotor
approximately 17 VA constant input

f(Hz)	rpm	input voltage (V)	input current (A)	phase lag (deg)	output torque (N m)
180.	3,550.	.387	24.	38.9	0.0
240.	4,950.	.398	22.	47.5	0.0
480.	10,050.	.532	16.	55.3	0.0
720.	15,800.	.661	13.	51.8	0.0
960.	20,400.	.755	11.	44.9	0.0
1,200.	22,100.	.841	10.	43.2	0.0

8 pole stator
squirrel cage rotor
approximately 17 VA constant input

f(Hz)	rpm	input voltage (V)	input current (A)	phase lag (deg)	output torque (N m)
240.	2,660.	.305	28.	17.3	0.0
480.	4,700.	.338	25.	27.6	0.0
720.	7,550.	.431	20.	23.3	0.0
1,200.	13,300.	.466	18.	30.2	0.0
1,440.	15,000.	.499	17.	31.1	0.0
1,920.	21,100.	.586	14.	24.2	0.0
2,400.	25,600	.627	13.	21.6	0.0

8 pole stator
steel jacketed squirrel cage rotor
approximately 17 VA constant input

f(Hz)	rpm	input voltage (V)	input current (A)	phase lag (deg)	output torque (N m)
240.	3,080.	.349	25.	38.9	0.0
480.	4,800.	.434	20.	48.4	0.0
720.	7,450.	.518	16.	41.5	0.0
1,200.	11,800.	.664	13.	38.9	0.0
1,440.	13,800.	.721	12.	36.3	0.0
1,920.	17,100.	.843	10.	34.6	0.0
2,400.	20,400	.910	9.	25.9	0.0

8 pole stator
solid rotor
approximately 17 VA constant input

f(Hz)	rpm	input voltage (V)	input current (A)	phase lag (deg)	output torque (N m)
240.	2,800.	.312	27.	21.6	0.0
480.	4,400.	.349	23.	34.6	0.0
720.	8,050.	.495	18.	25.9	0.0
1,200.	9,200.	.541	16.	30.2	0.0
1,440.	10,000.	.583	15.	25.9	0.0
1,920.	14,200.	.673	13.	27.6	0.0
2,400.	15,600	.756	12.	21.6	0.0

8.6.1.2 Blocked Rotor Tests.

4 pole stator
squirrel cage rotor
approximately 17 VA constant input

f(Hz)	rpm	input voltage (V)	input current (A)	phase lag (deg)
180.	0.0	.379	23.	32.4
240.	0.0	.392	22.	25.9
480.	0.0	.472	19.	20.7
720.	0.0	.520	17.	15.6
960.	0.0	.530	16.	6.9
1,200.	0.0	.553	16.	0.0

4 pole stator
laminated rotor with no squirrel cage
approximately 17 VA constant input

f(Hz)	rpm	input voltage (V)	input current (A)	phase lag (deg)
180.	0.0	.342	25.	38.9
240.	0.0	.356	23.	43.2
480.	0.0	.476	18.	51.8
720.	0.0	.584	15.	46.7
960.	0.0	.666	12.	41.5
1,200.	0.0	.774	11.	43.2

8.6.1.3 Load Tests.

4 pole stator
squirrel cage rotor
approximately 17 VA input at no load

f(Hz)	rpm	input voltage (V)	input current (A)	phase lag (deg)	output torque (N m)
240.	5,300.	.342	24.	38.9	0.0
240.	5,050.	.342	24.	38.9	.00022
240.	4,700.	.342	24.	38.9	.00044
240.	4,400.	.342	23.	34.6	.00065
240.	4,100.	.342	24.	34.6	.00087

4 pole stator
squirrel cage rotor
approximately 17 VA input at no load

f(Hz)	rpm	input voltage (V)	input current (A)	phase lag (deg)	output torque (N m)
720.	16,650.	.525	15.	51.8	0.0
720.	15,900.	.525	15.	46.7	.00022
720.	15,200.	.525	15.	41.5	.00044
720.	14,600.	.525	15.	36.3	.00065
720.	13,500.	.525	15.	36.3	.00087
720.	13,000.	.525	15.	36.3	.00108
720.	11,800.	.525	14.	31.1	.00131
720.	10,200.	.525	15.	31.1	.00153
720.	0.0	.525	19.	15.6	.0051

4 pole stator
squirrel cage rotor
approximately 17 VA input at no load

f(Hz)	rpm	input voltage (V)	input current (A)	phase lag (deg)	output torque (N m)
1200.	26,400.	.662	11.	38.9	0.0
1200.	24,500.	.662	12.	34.6	.00022
1200.	23,000.	.662	13.	30.2	.00044
1200.	20,900.	.662	13.	30.2	.00065
1200.	20,000.	.662	13.	25.9	.00087
1200.	18,200.	.662	13.	17.3	.00108
1200.	15,900.	.662	14.	17.3	.00131
1200.	12,100.	.662	14.	21.6	.00153
1200.	0.0	.662	21.		.0034

4 pole stator
steel jacketed squirrel cage rotor
approximately 17 VA input at no load

f(Hz)	rpm	input voltage (V)	input current (A)	phase lag (deg)	output torque (N m)
720.	16,500.	.845	10.	51.8	0.0
720.	15,600.	.845	11.	51.8	.00011
720.	14,650.	.845	11.	51.8	.00022
720.	12,450.	.845	12.	51.8	.00033
720.	9,250.	.845	12.	51.8	.00044
720.	6,700.	.845	12.	51.8	.00055
720.	0.0	.845	13.	51.8	.00127

4 pole stator
solid rotor
approximately 17 VA input at no load

f(Hz)	rpm	input voltage (V)	input current (A)	phase lag (deg)	output torque (N m)
720.	12,900.	.661	13.	51.8	0.0
720.	10,900.	.661	13.	46.7	.00007
720.	10,500.	.661	13.	46.7	.00011
720.	9,500.	.661	13.	46.7	.00017
720.	7,600.	.661	13.	46.7	.00022
720.	7,100.	.661	13.	46.7	.00028
720.	5,900.	.661	13.	46.7	.00033
720.	3,900.	.661	13.	46.7	.00039
720.	1,550.	.661	13.	46.7	.00044
720.	0.0	.661	13.	46.7	.00076

8.6.2 Liquid Nitrogen Temperature Tests

8.6.2.1 Blocked Rotor Tests.

4 pole stator
squirrel cage rotor
input currents are the same as in the room temperature tests

f(Hz)	rpm	input voltage (V)	input current (A)	phase lag (deg)
180.	0.0	.114	23.	32.4
240.	0.0	.124	22.	34.6
480.	0.0	.158	19.	34.6
720.	0.0	.204	17.	36.3
960.	0.0	.230	16.	31.1
1,200.	0.0	.280	16.	25.9

4 pole stator
laminated rotor with no squirrel cage
input currents are the same as in the room temperature tests

f(Hz)	rpm	input voltage (V)	input current (A)	phase lag (deg)
180.	0.0	.220	25.	64.8
240.	0.0	.268	23.	64.8
480.	0.0	.442	18.	58.8
720.	0.0	.565	15.	51.8
960.	0.0		12.	48.4
1,200.	0.0	.728	11.	43.2

8.6.2.2 <u>Load Tests</u>.

4 pole stator
squirrel cage rotor
approximately 17 VA input at no load

f(Hz)	rpm	input voltage (V)	input current (A)	phase lag (deg)	output torque (N m)
240.	5,550.	.297	30.	69.1	0.0
240	0.0	.297	46.	34.6	.00147

4 pole stator
squirrel cage rotor
approximately 17 VA input at no load

f(Hz)	rpm	input voltage (V)	input current (A)	phase lag (deg)	output torque (N m)
720.	16,650.	.525	15.	51.8	0.0
720.	15,900.	.525	15.	46.7	.00022
720.	15,200.	.525	15.	41.5	.00044
720.	14,600.	.525	15.	36.3	.00065
720.	13,500.	.525	15.	36.3	.00087
720.	13,000.	.525	15.	36.3	.00108
720.	11,800.	.525	14.	31.1	.00131
720.	10,200.	.525	15.	31.1	.00153
720.	0.0	.525	19.	15.6	.0051

4 pole stator
squirrel cage rotor
approximately 17 VA input at no load

f(Hz)	rpm	input voltage (V)	input current (A)	phase lag (deg)	output torque (N m)
1200.	26,400.	.662	11.	38.9	0.0
1200.	24,500.	.662	12.	34.6	.00022
1200.	23,000.	.662	13.	30.2	.00044
1200.	20,900.	.662	13.	30.2	.00065
1200.	20,000.	.662	13.	25.9	.00087
1200.	18,200.	.662	13.	17.3	.00108
1200.	15,900.	.662	14.	17.3	.00131
1200.	12,100.	.662	14.	21.6	.00153
1200.	0.0	.662	21.		.0034

9.0 OPTIMAL DESIGN FOR A SUPERCONDUCTING HIGH SPEED SYNCHRONOUS/INDUCTION MOTOR

A high T_c superconducting high speed synchronous/induction motor design is desired. Ideally this motor would be superior to a similarly sized normal induction motor. However as was pointed out in the conclusions of chapter 5.0 there are major obstacles.

A superconducting squirrel cage must use solid superconductors. There is no way to connect multifilament superconducting cable into superconducting junctions. This lowers the frequency withstand capability of the superconducting squirrel cage since the upper frequency limit of solid superconductor is below that of twisted and transposed multifilament superconductor cable.

A major shortcoming of both low and high T_c superconductors is their inability to superconduct when carrying significant currents of even moderate frequencies or experiencing significant magnetic fields of even moderate frequencies. Rechowicz[26] found a frequency limit of only about 1 Hz for the best low T_c superconductor's bundled filament cable in 1975. The frequency limit of solid low T_c large cross-section superconductors must be below that. Cave et al.[31] extrapolated from experimental data to predict AC losses in $YBa_2Cu_3O_x$ sintered superconductor. His predicted losses are comparable to those seen in low T_c

superconductors. See section 1.4.7 for more on the frequency limit.

As was determined in chapter 5.0, in a superconducting synchronous/induction motor the rotor superconductors are required to carry significant currents of 2 and 4 times the 60 Hz supply frequency. These frequencies are far above the capabilities of all present superconductors, low and high T_c.

A superconducting high speed synchronous/induction motor will need to use a supply frequency higher than 60 Hz. This means the frequencies experienced by the superconducting rotor bars will also have to be higher, again by the 2 and 4 times factors. For example a 2 pole 36,000 rpm superconducting synchronous/induction motor will need a supply frequency of 600 Hz. This means that the superconducting rotor bars must be able to carry significant currents and exist in significant magnetic fields of frequencies of 1,200 and 2,400 Hz. It is believed that the creation of superconducting materials that can withstand these frequencies is practically impossible.

Because of the frequency obstacle a design of a superconducting high speed synchronous/induction motor has not been done.

10.0 CONCLUSIONS

Conclusions have been made throughout this work. Chapters 1.0 to 3.0 and 8.0 include some conclusions within their texts. Chapters 4.0 to 7.0 have conclusion sections. This chapter will summarize or repeat the most important conclusions.

10.1 Superconducting Synchronous/Induction Motors

There are a number of serious problems with superconducting synchronous/induction motors.

1) The superconducting rotor bars must be able to withstand significant currents and magnetic fields of up to 4 times the stator supply frequency. With all present superconductors (low and high T_c) and a reasonable supply frequency, 60 Hz for instance, this is impossible.

Present superconductors, both low and high T_c, can only withstand current at 50 to 60 Hz when formed into cables of twisted filaments. In their solid form their frequency rating is much less. Twisted filament cables will not work in a squirrel cage, there is no way to make superconducting filament connections. Using present superconductors in their solid form the stator supply frequency must be less than 1/4th of the 1 Hz Rechowicz[26] gave for twisted AC cable in 1975. This means the maximum rotor speed will be less than 15 rpm, a speed too low for practical use.

2) Presently, high T_c superconductor materials, that can be made in large sizes, can not withstand current and magnetic flux densities of magnitudes necessary for reasonable superconducting synchronous/induction motor operation.

3) A high T_c superconducting squirrel cage is almost impossible to make. Available high T_c superconductor materials can not be welded or joined into superconducting junctions and are difficult to machine or mold into complex shapes.

4) Stability will be a problem with superconducting synchronous/induction motors. First, the desired trapped flux caused by the superconducting squirrel cage will cause the motor to act like a synchronous motor without damper bars. This was seen with the results of the two axis method analysis. Within the time limits of those results, the torque and speed oscillations didn't decay after transition. Second, undesirable trapped flux may cause instability.

5) The closeness of the rotor to the stator will cause stator loss heat to be transferred to the rotor. Even if the rotor losses are zero the stator magnetic and resistive power losses will transfer a significant amount of heat to the rotor. This heat will have to be countered with additional liquid nitrogen coolant. Because of the cost of

liquid nitrogen it is more economical to operate an ordinary room temperature induction motor than its liquid nitrogen cooled superconducting synchronous/induction motor equivalent.

Several observations were made as to how a superconducting synchronous/induction motor would operate, if one was made.

1) The power capacity and efficiency of a superconducting synchronous/induction motor would increase as its squirrel cage superconductor maximum current density increases, provided magnetic losses do not increase significantly.

2) The output power capacity of a superconducting synchronous/induction motor is bounded by the saturation of its ferromagnetic materials. Just as with induction motors, a superconducting rotor using standard laminations has a maximum flux density of about 2 T. The extremely high flux densities seen in a superconducting generator can not be used in a superconducting synchronous/induction motor.

3) Ignoring the cost of cooling, the efficiency of a superconducting synchronous/induction motor will improve over that of a similarly sized normal induction motor. In the example motor considered in the analysis done in chapters 4.0 to 6.0 it increased from .87 to .92 at rated output upon being converted to a superconducting

synchronous/induction motor. In chapter 7.0's analysis it increased from .87 to .93 at rated output upon being converted to a superconducting synchronous/induction motor.

10.2 High Speed Induction Motors

High speed induction motors do not have serious problems like superconducting synchronous/induction motors. They have been built and tested by many researchers.

1) As motor speed increases the supply frequency also must increase. This causes a significant increase in magnetic power losses in the stator laminations. If the frequency is too high the magnetic power losses will ruin the motor's efficiency and cause it to overheat. The use of low magnetic loss amorphous iron in the stator should be considered.

2) High speed induction motors should have two poles.

3) Laminated squirrel cage rotors should be used on high speed induction motors as they are on normal speed induction motors. Solid rotors create too many losses to make their simpler construction worthwhile.

4) Centrifugal forces and imbalance can destroy high speed rotors and bearings. Care must be taken in construction. Rotor bar slots should be overhung to better contain the rotor bars. Rotor fan blades should be reduced or eliminated.

5) Motor bearings are a major limitation to rotor

speed. Each bearing type has an upper speed limit. These limits are lower for larger diameter bearings due to the greater peripheral speeds. The limits are also lower for the more robust bearings such as ball bearings than for more unstable and smaller load capability bearings such as air or magnetic bearings.

6) Due to excessive magnetizing currents, resistive power losses in miniature high speed induction motors are much less efficient than similarly sized high speed permanent magnet synchronous motors.

11.0 SUGGESTIONS FOR FUTURE WORK

1) Much better superconductor materials are needed.

2) A method of making superconducting junctions is needed.

3) Stator magnetic materials should be improved on. Amorphous iron appears to have the potential for reducing magnetic losses at high frequencies.

4) Improved bearings are needed. Rugged high speed bearings would make high speed motors more useful in practical applications.

BIBLIOGRAPHY

1. Fink, Donald G. and Beaty, H. Wayne (ed.), Standard Handbook for Electrical Engineers (12th edition; New York: Mcgraw-Hill Book Company, 1987), p. 20.40.

2. Bugeat, J. P., Petit, R. and Valentian, D., "Electric Motor with Superconducting Winding," Cryogenics, Vol. 27, (February, 1987), pp. 59-62.

3. Bumby, J. R., Superconducting Rotating Electrical Machines (Oxford: Clarendon Press, 1983), p. 2.

4. Leonhard, A., "Asynchronous Machines with Superconductive Secondary Winding, Limiting Values of the Torques of Induction Machines in the Transient State," Arch. Elektrotech (Germany), Vol. 52, No. 4 (1969), pp. 205-210.

5. Brechna, H. and Kronig, H., "Three-phase Induction Motor with a Superconducting Cage Winding," IEEE Trans. Magn., Vol. MAG-15, No. 1 (Jan. 1979), pp. 715-718.

6. De Doncker, R. and Novotny, D. W., "High Temperature Superconductors in Electric Power Equipment?," Electric Machines and Power Systems, Vol. 13, No. 5 (1987), pp. 329-345.

7. Lipo, T. A., "The Potential for High Temperature Superconducting AC and DC Motors," Electric Machines and Power Systems, Vol. 13, No. 6 (1987), pp. 373-385.

8. Jordan, Howard. L., "Feasibility Study of Electric Motors Constructed with High Temperature Superconducting Materials," Electric Machines and Power Systems, Vol. 16, No. 1 (1989), pp. 15-23.

9. Nailen, R. L., "That Super Superconductor Motor: Is It Really a Breakthrough?," Electrical Apparatus, Vol. 41, (May, 1988), pp. 28-29.

10. Bumby, Op. Cit., pp. 63-146.

11. Nasar, S. A. (ed.), Handbook of Electrical Machines, "Superconducting Machines, by C. Flick" (New York: McGraw-Hill Book Company, 1987), Chapter 10.

12. "Super High-Speed Motor Rides on Air Cushion," Product Engineering, Vol. 42, (Jan. 18, 1971), p. 61.

13. Lakerveld, H. G., "High-Speed Solid-Rotor Induction Motors," Philips Tech. Rev., Vol. 34, No. 7 (1974) pp. 170-179.

14. Hesmondhalgh, D. E. and Tippings, D., "Miniature Sheet-Rotor Induction Motors for Extremely High Speeds," IEE Proc. B. Electrical Power Appl., Vol. 130, Pt. B, No. 4 (July, 1983), pp. 265-277.

15. Hesmondhalgh, D. E. and Tippings, D., "Some Problems Associated with Designing Motors for Speeds of up to 150000 RPM," Proc. Inter. Conf. on Electrical Machines, Swiss Federal Inst. Technol., Zurich, Switzerland, Vol. 1, (Sept., 1984), pp. 302-305.

16. Hesmondhalgh, D. E. and Tippings, D., "A Miniature Transformer-Coupled High-Speed Synchronous Motor," IEE Proc. B, Electr. Power Appl., Vol. 132, Pt. B, No. 6 (Nov., 1985), pp. 319-326.

17. Tanaka, H. and Kuga, T., "2000 kw-8000 rpm Very High Speed Induction Motor Drive System," IEEE IAS Annual Conference Record, (October, 1985), pp. 676-680.

18. Petit, P. and Delassus, J., "Utilisation des Moteurs Asynchrones á Rotor Massif dans les Machines á Vitesse Rapide," Techniques Cie Electro-Mechanique, No. 87 (June, 1973), pp. 22-26.

19. Ikeda, M., Sakabe, S. and Higashi, K., "Experimental Study of High Speed Induction Motor Varying Rotor Construction," IEEE Transactions on Energy Conversion, Vol. 5, No. 1 (March, 1990), pp. 98-103.

20. Fukao, T., Chiba, A. and Matsui, M., "Test Results on a Super-High-Speed Amorphous-Iron Reluctance Motor," IEEE Trans. Ind. Appl., Vol. 25, No. 1 (Jan./Feb., 1989), pp. 119-125.

21. Pouillange, J. P., "Moteurs Cylindriques à Aimants Permanents: de Bonnes Performances aux Moyennes et Grandes Vitesses," Revue Generale de l'Electricité, No. 3 (March, 1987), pp. 62-64.

22. Matsuda, I. and Mizuno, T., "High Speed Permanent-Magnet Type Synchronous Motor (PM Motor)," Meiden Review, Series No. 81, No. 3 (1987), pp. 36-38.

23. Miller, T. J. E., Brushless Permanent-Magnet and Reluctance Motor Drives, (Oxford: Clarendon Press, 1989) p 18.

24. Yoda, H., Kamata, K. and Shimizu, K., "Submerged Motor Pump for Cryogenic Plants," Hitachi Review, Vol. 32, No. 1 (1983), pp. 19-22.

25. Fitzgerald, K., "Superconductivity: Fact vs. Fancy," Spectrum, Vol. 25, No. 5 (May, 1988), pp. 30-41.

26. Rechowicz, M., Electric Power at Low Temperatures (Oxford: Clarendon Press, 1975), p. 51.

27. Mower, T. M. and Iwasa, Y., "Experimental Investigation of AC Losses in Cabled Superconductors," Cryogenics, Vol. 26, (May, 1986), pp. 281-292.

28. Yamamoto, M., Ikeda, M., Tanaka, Y., Matsumoto, K., Ishigohka, T. and Tsukamoto, O., "Development of 50 Hz Disc-Type Superconducting Coil," IEEE Trans. Magn., Vol. MAG-23, No. 2 (March, 1987), pp. 557-560.

29. Abell, J. S., Colclough, M. S., Forgan, E. M., Gough, C. E., Muirhead, C. M., O'Conner, D. A., Vinen, W. F., Wellhofer, F., Birchall, J. D., Alford, N. McN., Clegg, W. J. and Harmer, M., "Critical Currents, Magnetisation, AC-Susceptibility and AC-Losses in Ceramic Superconductors," Physica C (Netherlands), Vol. 153-155, Pt. 1 (1988), pp. 359-360.

30. Goldfarb, R. B., Clark, A. F., Panson, A. J. and Braginski, A. I., "AC Susceptibility Measurements Near the Critical Temperature of Y-Ba-Cu-O," Extended Abstracts, High Temperature Superconductors, Proceedings of Symposium S (Spring Meeting of the Materials Research Society, Society based in Pittsburgh, PA), (23-24 April, 1987), pp. 261-263.

31. Cave, J. R., Mautref, M., Agnoux, C., Leriche, A., and Février, A., "Electromagnetic Properties of Sintered YBaCuO Superconductors: Critical Current Densities, Transport Currents and AC Losses," Cryogenics (supplement), Vol. 29, (March, 1989), pp. 341-346.

32. Fisher, Greg, "Superconductor Mysteries Unravel as Developments Proceed," Ceramic Bulletin, Vol. 67, No. 4 (April, 1988), pp. 725-735.

33. Schlabach, T. D., "High-T_c Superconductor Wire Fabrication and Prospects for use in Large Scale Applications," Superconductivity Applications and Developments (ASME Materials Div.), Vol. 11, (Nov. 22-Dec. 2, 1988), pp. 29-34.

34. Smock, R. W., "Power Industry Eyes New Superconductors with Caution," Power Engineering, Vol. 91, No. 7 (July, 1987), pp. 34-37.

35. Sabrie, J. L. (Alsthom), "Moteurs Cryogéniques: L'Aube de l'Utilisation Industrielle," Revue Generale de l'Electricité, No. 3 (March, 1987), pp. 53-61.

36. Rechowicz, Op. Cit., p. 122.

37. Gershman, R. and Sherman, A. L., Fluid Properties Handbook (Santa Monica, CA: Douglas Aircraft Co., 1962), pp. 154.

38. Alger, P. L., Induction Machines (2nd edition; New York: Gordon and Breach Science Publishers, 1970), pp. 346-361.

39. Hamilton, H. B., "Electrical Design of an Induction Motor and Associated Investigations," (University of Minnesota: Minneapolis, Minn., 1954), unpublished.

40. Kuhlmann, J. H., Design of Electrical Apparatus (3rd edition; New York: John Wiley & Sons, Inc., 1950), pp. 317-320.

41. Kuhlmann, Op. Cit., pp. 289-355.

42. Kuhlmann, Op. Cit., p. 487.

43. Fitzgerald, A. E. and Kingsley, C., Electric Machinery (New York: McGraw-Hill Book Company, 1952), p. 422.

44. Brown, David and Hamilton, E. P. III, Electromechanical Energy Conversion (New York: Macmillan Pub. Co., 1984), pp. 100-109.

45. Fink, Op. Cit., p. 2.5.

Made in United States
North Haven, CT
05 November 2026

21760244R00135